# THE RELATIVISTIC BRAIN

# THE RELATIVISTIC BRAIN

## HOW IT WORKS AND WHY IT CANNOT BE SIMULATED BY A TURING MACHINE

Ronald Cicurel

Miguel Nicolelis

*Kios Press*          Natal, Montreux, Durham, São Paulo

Other books by the authors:

L'ordinateur ne digérera pas le cerveau: Science et cerveaux
artificiels. Essai sur la nature du reel.
by Ronald Cicurel

Beyond Boundaries: The new neuroscience of connecting brains
with machines and how it will change our lives
by Miguel Nicolelis

TED talks by Miguel Nicolelis:

1 Brain-to-brain communication has arrived.

2 A monkey that controls a robot with its thoughts.

Cover art and illustrations by Katie Zhuang

Kios Press, Natal, Montreux, Durham, São Paulo,
Copyright © 2015 by Ronald Cicurel and Miguel A. L. Nicolelis, 2015
ISBN 978-1511617024
(1.1)

To Wagner Weidebach and the fond memories that no machine
will ever bring back.

# Contents

# ACKNOWLEDGEMENTS

Many people contributed to the preparation and revision of this monograph. First, we would like to thank all the members of the Nicolelis Lab and the Edmond and Lily Safra International Institute of Neuroscience of Natal who were patient enough to read multiple versions of the manuscript and provide many suggestions, criticisms, and ideas to improve it. We are also grateful for all the comments forwarded to us by Dr. Ron Frostig, who helped us clarify our thoughts in many domains.

Katie Zhuang was in charge of the cover art and all illustrations included in this work. We are very much in debt for her great artistic touch. Susan Halkiotis revised the monograph many times and provided many insightful suggestions to improve the way we communicate complex ideas. All the experimental work from the Nicolelis' lab described in this monograph was supported by grants from the National Institutes of Health, FINEP, and the Hartwell Foundation

The authors would like to thank Lilliane Mancassola and M. Pitanga for their unconditional love and continuous support.

# PREFACE

This small volume attempts to summarize almost a full decade of discussions between a mathematician (RC) and a neurophysiologist (MN) who grew fascinated, in their distinct ways, about the brain and its never ending riddles and mysteries. Throughout the past 10 years, in long meetings that either took place or eventually ended in the Da Carlo pizzeria, located in the center of Lausanne, Switzerland, each of us first strove to learn the scientific language of the other and, after common ground was found, merge our best ideas into a single unabridged theory of the brain.

Although some of our individual ideas have appeared before in the form of two books ("Beyond Boundaries", written by MN, and L'ordinateur ne digérera pas le cerveau, written by RC), the present book contains the first comprehensive written report of these decade-long conversations over pizza Napolitana and Classic (RC) or Diet Coca-Cola (MN). As such, it represents our best effort to present such a complex topic in a language that can be accessible by multiple scientific communities, and by members of a lay audience eager to listen to such abstract conjectures.

Despite the fact that any reader will be able to understand the key claims made in this book, we should emphasize at the outset that further readings will be needed to obtain a full understanding – and thorough satisfaction - of the two central theses disclosed in this manuscript. The first of these is the description of a comprehensive new theory on how complex brains like ours actually operate. Named the relativistic brain theory (RBT), for reasons that we hope will become clear after the reader completes the first two chapters of this book, this new conceptual and physiological model tries to explain how huge networks of interconnected brain cells – known as neurons – are capable of both generating complex brain functions, ranging from mundane pain, to the sense of self and even consciousness, and, in striking contrast, diverge into pathological mental states, as the result of a multitude of neurological or psychiatric disorders that can devastate one's existence by altering the way our brains build the perception of ourselves and the world around us.

As a scientific theory, the RBT makes a series of predictions, which will require extensive experimental testing to be validated or falsified (see Appendix I). Nonetheless, we firmly believe that enough experimental and clinical evidence already available in the neuroscience literature warrants taking the opportunity to disclose our thoughts at this very moment. Indeed, to us, the relativistic brain theory offers a complete paradigm shift on how we comprehend the brain of higher animals.

In addition to disclosing a new brain theory, we also present in this monograph a series of arguments to counter the hypothesis, known as computationalism, that complex brains like ours resemble digital computers and, as such, could be reproduced or simulated by software running in a sophisticated supercomputer. For quite some time now, many computer scientists involved in artificial intelligence research have argued that most, if not all, complex functions generated by animal brains, including our own, will soon be effectively simulated by algorithms running on digital computers. Such a proposition is not completely new, since similar claims have been made since the days of Georges Boole and Alan Turing. Today, this computationalism view states that since the brain is a physical entity, it must obey the laws of physics which can be simulated on a digital computer.

More recently, a growing number of neuroscientists have advocated that the fulfillment of this epic milestone depends primarily on throwing bigger and faster computers and "big data" tools to the problem. Since vast funds (hundreds of millions of dollars) are required to acquire such hardware and manage the huge research teams involved in this effort, large research consortia have been formed around the globe with the explicit goal of simulating animal, and eventually even human, brains, on state-of-the-art supercomputers. Surprisingly, despite the enormity of such a claim – that a human brain will soon be simulated on a digital computer - and the potentially life-changing impact on human society that fulfillment of such an objective would bring, very little scientific scrutiny has been given to examining the very basic tenant of this proposal:

Can a digital machine simulate the higher functions of a human brain?

Here, we combine mathematical, computational, evolutionary and neurophysiological arguments to deny the feasibility of such claims. We argue that modeling organisms, such as animal brains, in digital computers is hindered by non-computable and non-tractable problems that not even modern supercomputers can effectively handle. On the contrary, using our relativistic view of the brain, we propose that complex central nervous systems generate, combine and store information about itself, the body and the external world through the recurrent dynamic interplay of a hybrid digital-analog computational engine. At this engine's core, the electrical firing produced by widely distributed networks of neurons flows through a large variety of "biological coils", formed by the nerve bundles of the brain's white matter core, continuously generating variable and complex electromagnetic fields (NEMFs). According to our theory, the manifold created by the continuous interferences of these NEMFs works as a "biological analog computer", the neuronal space-time continuum, from which a "mental space" emerges to underlie most of the brain's higher order functions. Interactions between this "mental space" and incoming peripheral sensory signals are stored in a distributed way on the brain. Since neither the generation of the NEMFs, nor their interplay with billions of neurons is either tractable or computable, any attempt to effectively simulate the true complexity of brains in a digital computer or any other Turing machine has no credible chance to succeed.

Natal, Montreux, Durham, São Paulo 2015

Ronald Cicurel
Miguel A. Nicolelis

# CHAPTER 1 – Listening to populations of neurons and building brain-machine interfaces: the experimental roadmap to probe a relativistic brain

Since the foundation of modern neuroscience, at the end of the XIX century, many generations of neuroscientists have entertained the idea that populations of neurons, instead of single brain cells, are responsible for generating all the unique behaviors and neurological functions that emerge from complex animal brains, including the human central nervous system (Hebb 1949). Yet, only during the past 25 years, thanks to the introduction of new neurophysiological and brain imaging techniques, has this hypothesis been tested extensively on a variety of animal and human studies (Nicolelis 2008).

Among the new approaches employed in animal studies, the method known as chronic, multi-site, multi-electrode recordings (CMMR) has provided the most comprehensive data in favor of the notion that populations of neurons define the true functional unit of the mammalian brain (Nicolelis 2008; Nicolelis 2011). Thanks to this neurophysiological method, tens to hundreds of hair-like, flexible metal filaments, known as microelectrodes, can be implanted in the brains of rodents and monkeys respectively (Schwarz, Lebedev et al. 2014). Basically, such microelectrodes serve as sensors that allow one to simultaneously record the electrical sparks – known as action potentials - produced by hundreds to thousands of individual neurons, distributed across multiple structures that define a particular neural circuit, like the motor system, which is responsible for generating the higher motor plan needed for producing limb movements. Because of the characteristics of the material used to produce these microelectrodes, these neuronal recordings can continue for many months or even several years (Schwarz, Lebedev et al. 2014).

About 15 years ago, one of us (MN) took advantage of the new possibilities opened by the introduction of the CMMR to create, together with John Chapin, then at Hahnemann University, a new experimental paradigm that was named brain-machine interfaces (BMIs) (Nicolelis 2001). In their original papers (Chapin, Moxon et al. 1999; Wessberg, Stambaugh et al. 2000; Carmena, Lebedev et al. 2003; Patil, Carmena et al. 2004) involving studies in rats and monkeys, Nicolelis and Chapin proposed that BMIs could serve as a very important tool to investigate the physiological principles governing how large populations of neurons interact in order to generate motor behaviors (Nicolelis 2003; Nicolelis and Lebedev 2009). Soon thereafter, in the early 2000s, the same authors proposed that BMIs could also provide a framework for developing a new generation of neuroprosthetic devices aimed at restoring movements in patients suffering from devastating levels of body paralysis, produced by either traumatic spinal cord injuries or as a consequence of a variety of neurodegenerative disorders (Chapin, Moxon et al. 1999; Nicolelis 2001; Nicolelis and Chapin 2002; Nicolelis 2003).

The potential of BMIs to provide new neurorehabilitation therapies was covered in two TED talks given by one of us (MN) over the past three years. In the first talk (http://tinyurl.com/n4pwx9p), on April 2012, the basic experiments that validated the feasibility of building operational BMIs in primates were presented. The talk also reviewed the plans to build a brain-controlled robotic vest, known as an exoskeleton that could be employed to restore lower limb mobility in paraplegic patients.

In a second TED talk, on October 2014 (http://tinyurl.com/lez9agu), the preliminary clinical results, obtained by 8 paraplegic patients who tested such an exoskeleton were described. Both the exoskeleton and the comprehensive neurorehabilitation protocol designed to train patients on how to use it were designed and implemented as a result of the collaboration of more than 100 scientists working as part of a non-profit, international research consortium, named the Walk Again Project.

After undergoing a multi-stage training program, these 8 patients not only learned to use their own brain activity, sampled through a non-invasive technique known as electroencephalography (EEG), to control the movements of the exoskeleton and walk again, but they also experienced vivid "phantom" tactile and proprioceptive sensations, emanating from legs they could not move or feel since the day they suffered a spinal cord injury. These "phantom" leg sensations were associated with the patients' newly reacquired locomotive ability. Indeed, above a certain exoskeleton speed, all patients reported the sensation of "walking by their own means", as if they were not being supported and helped by a robotic device. Such realistic sensations emerged as a result of the type of tactile feedback the patients received from arrays of pressure sensors distributed across the surface of the foot and joints of the exoskeleton. Thus, once the exoskeleton foot made contact with the ground, the pressure signal generated by the foot sensor was transmitted to an array of vibro-mechanical devices embedded in the sleeves of a "smart shirt" worn by the patient. Minutes after starting practice with this feedback system, patients reported phantom leg illusions, suggesting that the experimental apparatus could "fool" the brains into interpreting the vibration signals delivered to the skin of their forearm as if they were generated by their own biological feet and legs.

The patients became so proficient in using the first BMI-controlled exoskeleton that one of them, Juliano Pinto, paralyzed from the mid-chest down, was capable of delivering, on the sideline of a soccer field, the opening kick of the 2014 FIFA Soccer World Cup in Brazil.

While the World Cup demonstration of a brain-controlled exoskeleton unveiled to a larger audience the significant clinical potential that BMIs will have in the future, as predicted originally, BMI research has also generated a huge amount of experimental data related to how brain circuits operate in freely behaving animals. Altogether, these findings support a very different view on the physiological principles governing the cortex, the layered structure that underlies the most complex mental functions of the mammalian brain.

The key features of this new model of brain function were summarized in a series of principles of neural ensemble physiology (see Table 1), derived from the analysis of simultaneous recordings of the activity of 100-500 cortical neurons involved in the operation of different BMIs, created to investigate how limb movements are generated by the motor system. At the top of this list is the distributed principle, which states that all behaviors generated by complex animal brains like ours depend on the coordinated work of populations (or ensembles) of neurons, distributed across multiple brains structures.

Table 1 | **Principles of neural ensemble physiology**

| Principle | Explanation |
| --- | --- |
| Distributed coding | The representation of any behavioural parameter is distributed across many brain areas |
| Single-neuron insufficiency | Single neurons are limited in encoding a given parameter |
| Multitasking | A single neuron is informative of several behavioural parameters |
| Mass effect principle | A certain number of neurons in a population is needed for their information capacity to stabilize at a sufficiently high value |
| Degeneracy principle | The same behaviour can be produced by different neuronal assemblies |
| Plasticity | Neural ensemble function is crucially dependent on the capacity to plastically adapt to new behavioural tasks |
| Conservation of firing | The overall firing rates of an ensemble stay constant during the learning of a task |
| Context principle | The sensory responses of neural ensembles change according to the context of the stimulus |

*Table 1 – Principles of neural ensemble physiology. Published with permission from Nature Publishing, originally appeared in Nicolelis MAL, Lebedev MA. Principles of Neural Ensemble Physiology Underlying the Operation of Brain-Machine Interfaces. Nat. Rev. Neurosci. 10: 530-540, 2009.*

The distributed principle was clearly illustrated when monkeys were trained to employ a BMI to control the movements of a robotic arm using only their brain activity (see Figure 1.1A), without any overt movement of their own bodies. In these experiments, animals could only succeed when the combined electrical activity of a population of cortical neurons was fed into the BMI. Any attempt to use a single neuron as the source of the motor control signals to the BMI failed to produce the correct robot arm movements.

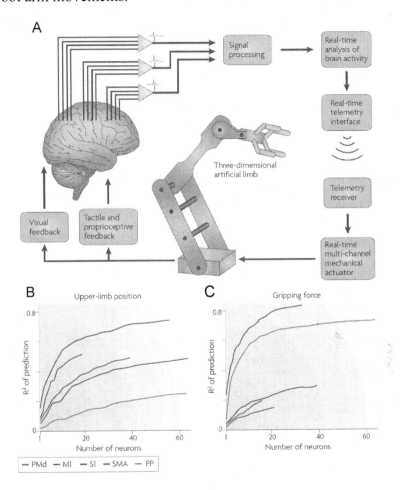

*Figure 1.1 –Principles of a brain–machine interface. (A) A schematic of a brain–machine interface (BMI) for reaching and grasping. Motor commands are extracted from cortical sensorimotor areas using multi-electrode implants that record neuronal discharges from large ensembles of cortical cells. Signal-processing algorithms convert neuronal spikes into the commands to a robotic device (e.g. arm, leg or wheel chair). Wireless telemetry can be used to link the BMI to the manipulator. The subject receives visual and somatosensory feedback from the actuator (B) Neuronal dropping curves for the prediction of arm movements in rhesus macaques calculated for populations of neurons*

*recorded in different cortical areas: the dorsal premotor cortex (PMd), the primary motor cortex (M1), the primary somatosensory cortex (s1), the supplementary motor area (sMA) and the posterior parietal cortex (PP). Neuronal dropping curves describe the accuracy (R2) of a BMI's performance as a function of the size of the neuronal population used to generate predictions. The best predictions were generated by the primary motor cortex (M1), but other areas carried significant information. Prediction accuracy improved with the increase in the number of recorded neurons. (C) Predictions of hand gripping force calculated from the activity of the same cortical areas as in panel A. Published with permission from Nature Publishing, originally appeared in Nicolelis MAL, Lebedev MA. Principles of Neural Ensemble Physiology Underlying the Operation of Brain-Machine Interfaces. Nat. Rev. Neurosci. 10: 530-540, 2009.*

Moreover, it was noticed that neurons distributed across multiple areas of the frontal and even parietal lobe, in both cerebral hemispheres, could contribute significantly to the population needed to execute the motor task. Further quantification of these results led to elucidation of yet another principle, the neural mass principle, which posits that the contribution (or prediction capability) of any population of cortical neurons to encode a behavioral parameter, like one of the motor parameters employed by our BMIs to generate robotic arm movements, grows as a function of the logarithm (base 10) of the number of neurons added to the population (Figure 1.1B). Since different cortical areas exhibited different levels of specialization, the slope of this logarithm relationship varied from region to region (Figure 1.1B). Yet, all these cortical areas could contribute some meaningful information to the final goal: move the robot arm.

The multitasking and degeneracy principles are next in Table 1. The multitasking principle indicates that the electrical activity generated by individual neurons can contribute to multiple neural ensembles simultaneously (Figure 1.1C). As such, individual neurons can participate in the computation of multiple functional or behavioral parameters at once. For instance, in the experiment described in the previous paradigm, cortical neurons could contribute to the generation of two distinct motor parameters at the same time, e.g. direction of arm movement and hand gripping force (Figure 1.1C).

Next in line, the neural degeneracy principle posits that a given behavioral outcome, let's say moving your arm to reach for a glass of water, can be produced, at different moments in time, by a distinct combination of neurons. In other words, multiple neural ensembles can yield the same behavioral outcome at different moments in time. In fact, some evidence suggests that the same combination of neurons is never repeated to produce the same movement.

The context principle states that the global internal state of a brain at a given moment in time determines how the brain is going to respond to a sensory stimulus or the need to produce a motor outcome. This implies that, during different internal states, the same brain can respond to an incoming stimulus – let's say, a touch on the subject's skin – in a completely distinct way (Figure 1.2). Put in a slightly different way, the context principle postulates that the brain has its "own point of view" and it applies it to make any decision regarding a novel event. By taking advantage of experiences accumulated throughout the subject's lifetime, the brain continuously reshapes and updates its internal "point of view" (Nicolelis, 2011), which can be interpreted as an internal model of the surrounding world's statistics and subject's own sense of self. Thus, before any encounter with a new event, let's say a new tactile stimulus, the brain expresses its own point of view, which is reflected, in neurophysiological terms, by the sudden appearance of widely distributed "anticipatory" neuronal electrical activity, across most of the cortex and related subcortical structures (Figure 1.3). The presence of such an expectation signal explains why we like to say that the "brain sees before it watches."

But how could a brain formed by such vast networks of intertwined neurons reshape itself so quickly, literally from moment to moment, throughout one's entire lifetime, to adjust its internal point of view, which it uses to scrutinize any new piece of world information it encounters? That exquisite property, which creates a profound and unassailable chasm between the mammalian brain and any digital computer, defines the plasticity principle: the ability of the brain to continuously adapt its micro-morphology and function in response to new experiences. Essentially, the brain is like an orchestra whose very instruments keep changing as a function of the music being produced.

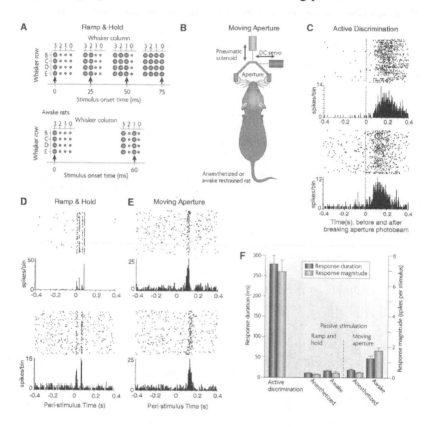

*Figure 1.2 – (A) The upper schematic C shows the pattern of multi-whisker ramp-and-hold passive stimuli delivered to anesthetized rats. Large black dots represent stimulation of a particular whisker. Upward arrows show stimulation onsets. The lower schematic shows the stimulation pattern of the awake restrained rats. (B) (Left) Schematic of the moving-aperture stimulus. The aperture is accelerated across the facial whiskers (with variable onsets and velocities) by the pneumatic solenoid and also simultaneously deflected laterally in varying amounts by the dc servo in order to accurately replicate the range of whisker deflection dynamics that occurred during active discrimination. (Right) Video frame captures showing an example of the aperture moving caudally across the whiskers of an awake restrained rat while simultaneously deflecting laterally 5 mm (to the right) over a 200-ms interval. (C) Representative single-unit responses showing long duration tonic*

*activation during active discrimination. The upper portion of each panel is a raster plot where each line represents a consecutive trial in a recording session, and each dot is a unit spike; the lower portion of each panel shows summed activity for all trials in 5-ms bins. The 0 time point represents the moment when rats disrupted the aperture photobeam. (D) Representative single- unit responses evoked by passive ramp-and-hold stimulation of 16 whiskers in lightly anesthetized rats (upper panel) and by passive stimulation of 8 whiskers in awake restrained rats (lower panel). The 0 time point represents stimulus onset. (E) Representative single-unit responses evoked by moving-aperture stimulation of awake restrained rats (the 0 time point represents the onset of aperture movement). (F) Mean (+SEM) excitatory response duration and magnitude evoked during active discrimination and by the different passive stimuli delivered to anesthetized or awake restrained rats. Permission requested, modified from Krupa DJ, Wiest, MC, Laubach M, Nicolelis MAL Layer specific somatosensory cortical activation during active tactile discrimination  Science 304: 1989-1992, 2004.*

According to the plasticity principle, the internal brain representation of the world, and even our own sense of self, remains in continuous flux throughout our lives. It is because of this principle that we maintain our ability to learn throughout life. Plasticity also explains why in blind patients, we can detect neurons in the visual cortex that respond to touch. That may explain why blind patients become so exquisitely proficient in reading Braille signals with their fingertips.

Recently, Eric Thomson and others in the Nicolelis Lab have shown how far the plasticity principle can be exploited in order to induce a piece of cortex to adapt to new outside world conditions (Thomson, Carra et al. 2013). By attaching an infrared detector to the frontal bone of adult rats and delivering the electrical output of this sensor directly to the region of the somatosensory cortex that processes tactile information generated by stimulation of the rat's facial whiskers, these researchers have induced these animals to learn how to "touch" otherwise invisible light (Thomson, Carra et al. 2013). As it is well known, mammals do not have retinal photoreceptors that are able to detect infrared light. Therefore, they are blind to infrared beams and cannot track them. After a few weeks of training with the apparatus created by Thomson and his colleagues, rats became very proficient in tracking infrared beams that led them to locations in which a reward could be acquired.

*Figure 1.3 – Ranking of neuronal ensembles reveals extensive anticipatory firing activity in primary motor (M1), primary somatosensory (S1) cortices, and the ventral medial (VPM), and posterior medial (POM) nuclei of the thalamus. (A) Peri-stimulus time histograms (PSTHs) of all areas studied showing different periods of increased or decreased neuronal firing activity spanning the whole duration of a task trial. Time 0 corresponds to the discrimination bar beam break. Neurons are not from the same animal. The top neuron was recorded in M1 and presented a period of increased firing activity only before the trial started. As soon as the door opened, this neuron decreased its activity. The onset of this decreased activity matched the beginning of firing increases observed in other M1 and in S1 neurons (second to fourth rows). This suggests an initial role for M1 at the preparatory stages of a trial, followed by a second class of cells both in M1 and S1 related to early anticipatory activity as the door opens (approximately - 0.5 s). As the animal moved from the door to the discrimination bars, anticipatory firing activity was observed in VPM and POM neurons. A M1 neuron (fifth through eighth rows) exhibited a sharp increase in firing activity that ended as the whiskers contacted the bars (time 0). Although not shown in this*

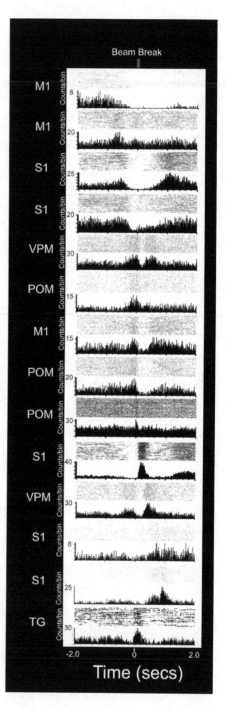

*figure, neurons with anticipatory increases of firing rate were present in all structures recorded. As this group of anticipatory cells decreased its firing, a different group of neurons in POM, M1, and S1 (9th through 11th rows) increased their activity. This period coincides with the whiskers sampling the discrimination bars. Also, as the whiskers touch the center nose poke and the rat chooses one of the reward ports (12th and 13th rows), firing increases were observed both in VPM and S1. Notice that after the whiskers had sampled the discrimination bars, increases of firing activity started to appear again in some of the upper rows neurons, suggesting that their activity was temporarily inhibited during tactile discrimination. On the bottom row, the activity of a typical TG neuron is presented. Between the door and the discrimination bars (~250 ms), there is almost no activity in this neuron, indicating that no whisker contacts or movements were made. A clear increase in TG activity is observed as the whiskers make contact with the tactile discriminanda. Overall, the combined PSTHs presented here show that active tactile discrimination results from complex interactions where all regions are likely to have a significant contribution at every point in time, and not just during a specific epoch (e.g., motor or tactile periods). Permission requested, originally published in Vieira M, Lebedev MA, Nicolelis MAL. Top-down Modulation in Cortico-Thalamo-Cortical Loops during Active Tactile Discrimination. J. Neurosci. 33:4076–4093, 2013.*

Because the IR tracking was done through the animal's somatosensory cortex, the authors proposed that rats experienced IR light as some sort of tactile stimulus. In support of this hypothesis, these investigators showed that, as training with the IR detector progressed, more and more individual neurons located in the somatosensory cortex of these rats became responsive to IR light. Yet, these neurons remained capable of responding normally to the mechanical displacement of their facial whiskers. Essentially, Thomson et al. were able to induce a piece of cortex to process a new sensory modality – infrared light – without inducing any reduction in normal tactile capacity.

Multi-electrode recording experiments in freely behaving rodents and monkeys have also revealed that, despite the continuous variation in an individual neuron's firing rate observed as these animals learned to perform a variety of behavioral tasks, the global electrical activity produced by large cortical circuits tends to remain constant. In other words, the total number of action potentials produced by a pseudo-random sample containing hundreds of neurons that belong to a given circuit – let's say the motor or somatosensory cortex – tend to hover tightly around a mean. This finding, which has been validated by recordings obtained from multiple cortical areas in several animal species (mice, rats, and monkeys), led to the formulation of the conservation of energy principle. This principle proposes that, due to the limited and constant energy budget available for the brain, neural circuits have to maintain a firing rate cap. Thus, if some cortical neurons increase their instantaneous firing rate to signal a particular sensory stimulus or to participate in the generation of a movement or other behavior, other neighboring cells will have to reduce their firing rate proportionally, so that the overall activity of the entire neural ensemble remains constant.

Although a few other principles have been derived from 25 years of multi-electrode experiments, the list reviewed above is sufficient to portray the kind of dilemma facing neuroscientists who seek to find some synthetic theory on how complex animal brains operate. Certainly, none of the classical theories of mainstream neuroscience could explain the findings that emerged from the multi-electrode recording experiments reported above. For starters, most of these theories do not take into account any notion of brain dynamics; from the millisecond scale, in which neural circuits operate, to the temporal scale in which brain plasticity occurs, brain dynamics has been utterly ignored for almost a full century of brain research. Thus, both the concept of time and the various manifestations of neuronal timing were never part of the classical central dogma of neuroscience, which remained dominated by static concepts such as cortical columns, maps and the never ending cataloguing of particular neuronal tuning properties.

In 2011, one of us (MN) published a book containing the early attempt to formulate a theory of brain function that could encompass all the major findings derived from neural ensemble recording studies in behaving animals (Nicolelis 2011). For started, this theory proposed that:

*"...from a physiological point of view, and in direct contrast to the classical twenty-century canon of cortical*

*neuroanatomy, there are no absolute or fixed spatial borders between cortical areas that dictate or constrain the functional operation of the cortex as a whole. Instead, the cortex should be considered as a formidable, but finite, neuronal space-time continuum. In this continuum, neurological functions and behaviors are allocated or produced respectively by recruiting chunks of neuronal space-time, according to a series of constraints, among which are the evolutionary history of the species, the layout of the brain determined by genetic and early development, the state of sensory periphery, the state of the internal brain dynamics, other body constrains, task context, the total amount of energy available to the brain, and the maximum speed of neuronal firing."*

By itself, the concept of the neuronal space-time continuum as a way to explain how the mammalian cortex operates was already a big jump. Yet, this central idea led to the formulation of a more comprehensive theory, which for reasons that will become apparent in Chapter 2, was named the Relativistic Brain Theory. According to the original formulation of this theory:

*"…when faced with new ways to obtain information about the statistics of the surrounding world, a subject's brain will readily assimilate those statistics, as well as the sensors or tools utilized to gather them. As a result, the brain will generate a new model of the world, a new simulation of the subject's body, and a new set of boundaries or constraints that define the individual's perception of reality and sense of self. This new brain model will then continue to be tested and reshaped throughout the subject's life. Since the total amount of energy the brain consumes and the maximal velocity of neuronal firing are both fixed, it appears that neuronal space and time would have to be relativized according to these constraints."*

But how does this neuronal space-time continuum emerge? What is the "glue" that keeps is working? What is the anatomical basis for supporting such a functional construct? What neurophysiological and behavioral phenomena can be better explained by this new construct? What are the key predictions made by the theory that can be used to falsify or validate it?

For the past 4 years, the two authors of this monograph have been engaged in discussing and expanding the original version of the relativistic brain theory. Part of this work involved seeking answers for the very questions raised in the previous paragraph. A summary of the outcome of this collaboration is the central theme of Chapter 2.

## CHAPTER 2 – The relativistic brain theory: its principles and predictions

The relativistic brain theory is a falsifiable scientific theory, which makes a series of testable predictions (see Appendix 1) about how the brains of mammals, and in particular those of primates and humans, work. The theory also provides novel physiological mechanisms to account for many unexplained experimental findings in animals and neurological observations in human subjects. Moreover, the theory suggests various new research directions at the interface of neuroscience and computer science.

According to the relativistic brain theory, complex central nervous systems like ours generate, process, and store information through the recursive interaction of a hybrid digital-analog computation engine (HDACE). In the HDACE, the digital component is defined by the spikes produced by neural networks distributed all over the brain, whereas the analog component is represented by the superimposition of time-varying, neuronal electromagnetic fields (NEMFs)[1], generated by the flow of neuronal electrical signals through the multitude of local and distributed loops of white matter that exist in the mammalian brain.

The relativistic brain theory proposes that the combination of time-varying NEMFs provides the "physiological glue" for the creation of the neuronal space-time continuum" in higher mammals. This space-time continuum defines what we call the "mental space, the analog neuronal substrate from which all higher brain functions emerge. Figure 2.1 illustrates the operation of this hybrid digital-analog computational engine

---

[1] This kind of analog computation was studied in the 1950's and named field computation. Interacting fields can produce summations by linear superposition, convolutions, Fourier transforms, wavelet transforms, Laplacian, etc. An EM field comprises an uncountable infinity of points. A such, it cannot be processed in a finite number of discrete steps.

(HDACE) as a highly recursive system in which NEMFs, generated by neuronal electrical activity, also influence, by induction, the electrical sparking of the very neuronal networks that participated in their creation, pretty much like, but on a different scale, the way a solar magnetic disturbance induces lightning storms on Earth.

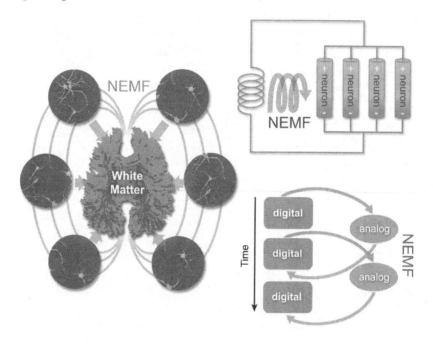

*Figure 2.1 – Multiple representation of the Hybrid Digital and Analog Computation Engine proposed by the Relativistic Brain Theory. (A) Distributed groups of neurons produce electrical signals which are transmitted through a vast network of neural fibers, collectively known as the white matter. This defines the digital component of the HDACE. As these electrical signals flow through the white matter, it generates a complex manifold of neuronal electromagnetic field (NEMF), which defines the analog component of the HDACE. The NEMF then influences, by induction, the behavioral of the pools of neurons that gave rise to it. The same concept is showed in (B) using an electrical circuit equivalent, where groups of neurons work as batteries and a coil generates the NEMF that acts upon the original groups of neurons. In (C) a block diagram represents the dynamic and recursive nature of the HDACE by showing that once groups of neurons (digital component) generate an NEMF (analog component), the latter will*

*influence the same group of neurons at a different time epoch, which defines a distinct internal brain state. The neurons, on their turn, generate a new NEMF, allowing the recursive process to continue.*

We should emphasize at this point that although we recognize the existence of other analog signals in the brain (e.g. membrane and synaptic potentials), here we are interested in analog signals capable of generating widespread emergent brain properties, rather than local phenomena[2].

Overall, the HDACE defines an integrated computing system. Accordingly, NEMFs would represent the actual materialization of the emergent neural properties postulated by neurobiologists to account for most of the higher brain functions, including things like our ability to experience pain sensations, our sense of self, and even our consciousness. NEMFs would also potentiate the brain's ability to use information from the outside world (and from inside the body) to reshape its neuronal circuits, a property known as causal efficiency, which manifests itself through the plasticity principle. Such continuous reshaping of the brain is vital for maintaining the kind of local negative entropy needed for the preservation of living animals.

From a neurophysiological point of view, there are many pieces of experimental and clinical evidence that are consistent with both the existence of widespread neuronal NEMFs and their potential role in brain processing, as proposed by our relativistic brain theory. For starters, electrical cortical fields have been measured since the mid-1920s through a technique known as electroencephalography (Berger 1929). Brain magnetic fields have also been measured through another method known as magnetoencephalography (Papanicolaou 2009). Those latter measurements, however, have been mainly confined to cortical NEMFs. Our theory, on the other hand, predicts the existence of widespread subcortical NEMFS too. Albeit small individually, these NEMFs could be generated through the innumerous bundles of axons that connect cortical and subcortical structures.

---

[2] HDACE does not exclude other computational levels within the components.

Figure 2.2 depicts an image of the human brain obtained with an imaging method called diffusion tension technique. This image illustrates how some of the bundles of axons that form the brain's white matter can define a true myriad of biological coils, from which very complex NEMFs would emerge.

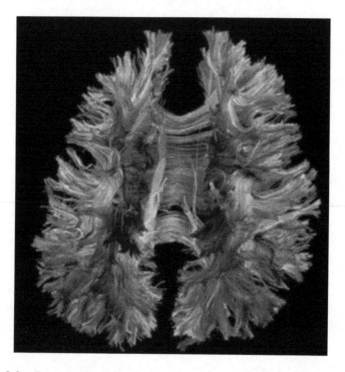

*Figure 2.2 – Primary data images containing tractography – obtained through diffusion tension imaging - from a representative normal human subject. Published with permission from Elsevier Publishing, originally appeared in Englander ZA, et al. Diffuse reduction of white matter connectivity in cerebral palsy with specific vulnerability of long range fiber tracts. NeuroImage: Clinical 2:440–447, 2013.*

Although experimental evidence for such deep subcortical NEMFs is lacking at the moment, indirect clues suggest that they, combined with cortical NEMFs, may exert an important role in creating widespread brain synchronous and coherent activity and, hence, lead to the emergence of the neuronal space-time continuum (Anastassiou, Montgomery et al. 2010)[3]. Consistent with this scenario, for the past 3 decades multiple laboratories have documented the existence of widespread neuronal synchronization in the gamma frequency band throughout the visual system (Engel, Fries et al. 2001). Indeed, this observation was proposed as a possible solution to the so called "binding problem" in perception (von der Malsburg 1995). The relativistic brain theory provides a solution for the "binding problem" by proposing that the brain does not need to reconstruct an original outside image of the world, sampled by the eyes, by binding "a posteriori" (or after the stimulus) one feature (e.g. color, orientation, shape) of the image after another at cortical level (von der Malsburg 1995). Instead, our theory proposes that, at any given moment in time, the brain generates its own internal analog hypothesis of what "it expects to see" by building an analog neural computer ahead of any encounter with a new visual stimulus.

To distinguish this operation from the classic binding problem, we shall refer to this phenomenon as the "a priori expectation" of a relativistic brain. In this new computational paradigm, incoming peripheral signals interfere with the internal analog engine that defines the "a priori expectation" created by the brain, modifying its final pattern, until the resulting image emerges in our minds. This mode of operation is not only very distinct from the solutions so far proposed for the binding problem", but, more importantly, it is incompatible with a computationalist digital model of the brain (Copeland 1998), which originally gave birth to the "binding" problem in the first place.

In a nutshell, the original conception of the binding problem only made sense because the feed-forward model of perception, put forward by David Hubel's and Torsten Wiesel's classical studies on the visual system (Hubel 1995), required the

---

[3] These authors have shown the effect of spatially inhomogeneous extracellular electric fields on neurons. They also showed that NEMFs generated by the cooperative action of brain cells can influence the timing of neuronal activity.

existence of some sort of syntactic representation system to allow perception of a complex visual scene to occur. Such a requirement is totally eliminated by the relativistic brain theory, rendering the binding problem simply moot.

High levels of anticipatory firing activity have also been observed, throughout the mammalian neocortex and thalamus, in rats trained to perform a tactile discrimination task (Pais-Vieira, Lebedev et al. 2013). This widespread build up in neuronal electrical activity appears a few hundred milliseconds before the animal engages its facial whiskers to touch an object (See Fig 1.3 in Chapter 1). Such synchronous electrical activity involves virtually the entire neocortex, spreading from the prefrontal, to motor, somatosensory, posterior parietal and all the way to the rat primary visual cortex. Moreover, all thalamic nuclei involved in tactile information processing exhibit the same type of anticipatory firing observed at cortical level (Pais-Vieira, Lebedev et al. 2013). Interestingly, disruption of this anticipatory synchronous activity leads to a significant decrease in animal performance in a tactile discrimination task (Pais-Vieira, Lebedev et. al. 2013). For the relativistic brain theory, this widespread anticipatory neuronal firing represents the manifestation of the "brain's own point of view", the internal world model built over the individual's lifetime which, at each new behavioral event or stimulus encounter, manifests itself as a series of expectations of what the animal may encounter (Nicolelis 2011).

Even in awake rats that are not engaged in a task, but are otherwise standing in an attentive posture, disseminated synchronous oscillatory activity can be measured in both cortical and subcortical structures that define the somatosensory system (Nicolelis, Baccala et al. 1995). These 7-12Hz bouts of synchronous firing originate in the somatosensory and motor cortices and soon after spread down to multiple thalamic nuclei and even brainstem structures (Figure 2.3).

By introducing the concept of the "brain's own point of view" as one of its central pillars, the relativistic brain theory also provides a physiological explanation for the findings that led to the formulation of the context principle discussed in Chapter 1.

According to this principle, the same sensory input – a visual cue or a tactile stimulus – may be signaled in a very distinct way by individual neurons or neural ensembles depending on whether animals are anesthetized, awake but immobile or fully engaged in actively searching the surround environment (see Figure 1.2

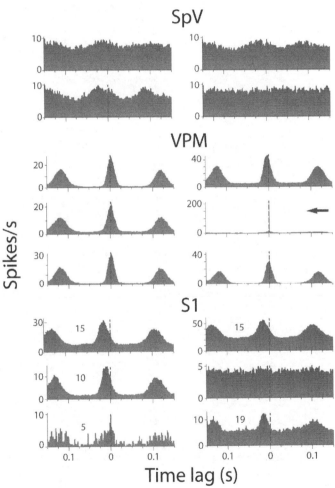

chapter 1).

*Figure 2.3 – Spontaneous 7- to 12 – Hz oscillations at multiple relays of the rat trigeminal somatosensory system. (A) Cross correlograms (CCs), calculated for 16 out of a total 48 simultaneously recorded neurons, reveal synchronous 7- to 12-Hz oscillations at three levels of the trigeminal pathway (spinal nucleus of the trigeminal complex (SpV), 4 neurons; ventral posterior medial (VPM) nucleus of the thalamus, 6 neurons; and primary somatosensory cortex (SI), 6 neurons). All CCs centered around the spiking of one reference VPM neuron (autocorrelation shown at arrow). Numbers on top of CCs indicate the time interval (in milliseconds) by which each SpV or SI neuron phase-leads the reference VPM neuron. All horizontal axes indicate pre- and post-VPM spike times (in seconds). VPM spiking occurred at 0.0s, as indicated by the vertical dashed lines. Vertical axes o CCs indicate equivalent firing rates per bin (bins: 1-ms duration). Notice that S1 neurons phase lead the VPM by at least 12ms. Permission requested, modified from Nicolelis MAL, Baccala LA, Lin RCS, Chapin JK. Sensorimotor encoding by synchronous neural ensemble activity at multiple levels of the somatosensory system. Science 268: 1353-1358, 1995.*

The relativistic theory proposes that this happens because, under each of these conditions, the internal dynamic state of the brain is different. As such, the manifestation of the "brain's own point of view" varies dramatically from an anesthetized animal (where it collapses to a point attractor) to a subject fully engaged in sampling its immediate whereabouts (where the brain's own point of view is fully expressed). Since the brain's response to the same sensory stimulus, at each moment in time, depends on the interference of the incoming sensory volley from the periphery with the brain's internal model of the world, neuronal sensory evoked responses should vary dramatically from anesthesia to fully awake/mobile conditions. That is precisely what has now been observed in a variety of experiments involving the tactile, gustatory, auditory and visual systems.

Extensive clinical findings also support the existence of an internal analog component of brain processing. For example, an interesting set of phenomena, known collectively as alterations of the body schema, are consistent with our relativistic brain theory and the existence of a HDACE. The most well-known of these illusions is the phantom limb sensation. This refers to the ubiquitous finding that patients who suffer the loss of a limb tend to experience its presence after such an amputation. Most amputees not only feel the presence of the missing limb, but they also report the presence of excruciating pain on a limb that does not exist any longer (Melzack 1973).

Another example of a body schema alteration is provided by the rubber hand illusion in which normal subjects report that a mannequin's hand feels like their own biological hand (Botvinick and Cohen 1998). This illusion is produced by first occluding one of the subject's hands from his/her view and then placing a mannequin hand/arm in front of the subject. Next, both the subject's occluded hand and the mannequin's hand are touched synchronously for a period of 3-5 minutes by the experimenter. When the experimenter stops touching the occluded subject's hand, but continues to touch only the mannequin's hand, most subjects experience an immediate illusion that the mannequin's hand feels like their real hand (Botvinick and Cohen 1998).

Recent neurophysiological experiments in monkeys, using a similar rubber hand paradigm, have shown that, just a few seconds after the animal's own hand stops being touched, but an avatar hand, projected on a computer screen in front of the animal's view, continues to be stimulated by a virtual probe, more than one-third of the neurons in the somatosensory cortex of the monkey continue responding to the visual stimulus provided by the virtual touches on the avatar arm as if the animal's own biological arm was being touched (Shokur, O'Doherty et al. 2013).

Both the phantom limb sensation and the rubber hand illusion suggest that the brain contains an "a priori" internal and continuous "body image" that can be reshaped very quickly as a function of the subject's experience. Ronald Melzack named this "body image" the neuromatrix (Melzack 1999), but did not elaborate on its physiological substrate.

The body image is a key component of what is known as the sense of self. According to the relativistic brain theory, the HDACE would easily provide a neurophysiological mechanism to explain these phenomena by proposing that the sense of self and the body image arise from a widely distributed NEMF generated by the combined electrical activity of many cortical and subcortical structures involved in the definition of the brain's body schema. Since no afferent body input can be generated by either an amputated limb or touching a mannequin 's hand, the body schema (and the sense of self) can only be described as a brain-derived expectation - an analog abstraction - of the subject's own body configuration. Thus, because the brain internally generates an expectation of what the subject's body should contain, based on genetic endowment and its stored memories, e.g. two arms, even when one of the subject's arms is amputated or occluded from vision, the sensation of this arm's existence continues to be experienced by the subject.

Further indirect evidence that our sense of self may emerge from hybrid digital-analog interactions, mediated by distributed cortical neural networks and cortical NEMFs, was obtained by the demonstration that transcranial magnetic stimulation (TMS) can modulate the phenomenon of "proprioceptive shift" observed during the rubber hand illusion (Tsakiris, Costantini et al. 2008). Proprioceptive shift refers to the subject's sensation that the position of his own hand shifts toward the position of the mannequin hand during the rubber hand illusion. If disruptive TMS is applied to the right temporal-parietal junction, 350 msec after induction of the rubber hand illusion, patients report a decrease in proprioceptive shift as compared to no-TMS trials (Tsakiris, Costantini et al. 2008). Furthermore, TMS applied to the visual cortex of blind patients can induce normal or abnormal tactile sensations, such as finger paresthesias. If these patients are engaged in Braille reading, TMS of the visual cortex can also impair their ability to read the embossed characters (Cohen, Celnik et al. 1997; Kupers, Pappens et al. 2007).

Another example that illustrates well our view on how complex mental experiences can be generated by the interaction of NEMFs that define the mental space is the sensation of pain. Although neurons related to different aspects of nociceptive information processing have been identified, how a complex integrated sensation of pain emerges from this distributed neural circuitry remains elusive. For example, further proof of the non-locality characteristics of pain can be found in the observation that it is not possible to elicit pain by electrically stimulating any part of the neocortex. According to the relativistic brain theory, such difficulty in pinpointing a precise source for the pain sensation results from the fact that pain, as any other complex mental property, emerges as a result of widely distributed interactions of NEMFs generated by the collective electrical activity of multiple cortical and subcortical circuits. In our relativistic terminology, the sensation of pain is the result of a multi-dimensional folding of the mental space, allowing multiple physiological factors (e.g. location, intensity, mnemonic references, emotional content, etc.) to be combined in order to shape the neuronal space-time continuum in a way that allows for the final pain percept to be experienced. Thus, by assuming that pain emerges in the analog component of the HDACE, as a result of a widespread combination of neural digital signals and mnemonic traces that combine to generate particular NEMFs, one could more easily identify a mechanism through which a subject's emotional, contextual, historical factors play such an important role in modulating incoming nociceptive signals from the body periphery in order to define the full qualia of a given pain episode.

The relativistic brain theory also proposes that the "mental space" is in charge of writing the result of its computational activity, performed in the analog domain, on the organic substrate that generates the digital component of the brain: the neural circuits. This idea is consistent with the notion that memories get consolidated in the neocortex during episodes of rapid eye movement sleep (REM) (Ribeiro, Gervasoni et al. 2004), a period of the sleep cycle characterized by dreaming and the mental replay of experiences acquired during the previous state of wakefulness. In our view, dreams would constitute examples of analog computer engines created with the purpose, among others, of inducing the reorganization and long-term storage of memories at the cortical level.

Several factors would constrain the shape and dynamics of this "mental space". Among others, we can cite the particular spatial distribution and composition of the neuronal pools in the brain, the structural features of the nerve pathways and loops of the white matter that connect these neuronal clusters, the energy consumption budget available for the brain, the neurotransmitters utilized by the nervous tissue, as well as the realm of previous mental experiences stored in the format of memories. Yet, despite all the complexity that emerges from this set of constraints, in our view this "mental space" could be amenable to mathematical analysis. Indeed, we propose that the geometry and topology of the "mental space", which defines a continuum multi-dimensional manifold, could be, in theory, investigated formally. Hints of the kind of mathematical approach that would be amenable to analyze the "mental space" derive from the observation that this internal neuronal space-time continuum can "curve" during a large range of mental experiences, like when we dream, experience illusions or suffer from hallucinations, generated by consumption of drugs (e.g. LSD) or mental disorders (e.g. schizophrenia). Curving of the mental space would explain why, in all these mental experiences, the notion of space, time, and even our own sensations (e.g. touch, vision, and temperature), body movements and sense of self can be warped. For instance, a person under the influence of LSD may think that the sidewalk in which he/she is walking is made of water and, therefore, may try to dive into hard concrete. This warping of space and time suggests to us that the geometry governing the "mental space" is rather Riemannian than Euclidean. As such, a particular type of algebra could be employed to analyze the type of mental experiences underlying the learning of new concepts and behaviors, as well as the neuronal etching of new memories by NEMFs acting on neuronal networks.

As far as we can say, this is the first time the idea of higher brain functions as emergent properties resulting from time-varying NEMFs has been given a potential physical definition and a possible mathematical description, through the analysis of the postulated geometry and topology of the "mental space". Thus, the relativistic brain theory, if validated by extensive experimentation, offers a radically different view on both normal and pathological brain operation.

The relativistic brain theory also provides for a biological mechanism that can generate useful abstractions and generalizations quickly, something that a digital system would expend a lot of time trying to mimic. Furthermore, NEMFs that account for a given behavioral outcome can be generated by different combinations of neuronal elements, at different moments in time, providing support for the neuronal degeneracy principle described in Chapter 1. In other words, different neuronal ensembles can produce a similar analog computational brain engine at different time epochs. By the same token, in the HDACE, different groups of neurons could be recruited by NEMFs to store the outcome of a given internal analog computation. That would be consistent with the current notion that long-term memories are stored in a distributed fashion across the cortical tissue. Indeed, without the existence of the analog brain component, it would be very difficult to explain how cortical circuits, characterized by complex micro-connectivity is continuously changing, could recall the type of precise information needed for memories to emerge virtually instantaneously throughout (most of) one's life.

In summary, the relativistic brain theory predicts that key non-computable processes like perception, mental imagery, and memory recall occur on the analog domain, thanks to the emergence of analog computational engines formed by time-varying NEMFs. The combination of these NEMFs, therefore, would define what we call the "mental space", a space in which all relevant mental activities of a brain take place throughout the subject's life. The content of this mental space is essentially what a human being describes when he is speaking about himself and his worldview.

Altogether, the existence of an analog domain also endows the animal brain with yet another level of plastic adaptation capability. That would also explain a multitude of recent findings which suggest that most of the neocortex is of a multi-sensory nature (Ghazanfar and Schroeder 2006; Nicolelis 2011). Indeed, if one assumes that NEMFs can fuse neuronal space and time to form a continuum at the cortical level, as the result of "a priori expectations", in theory, any part of the cortex could be recruited for mediating, at least partially, a particular high demand task in the "mental space". That would explain, for instance, why the visual cortex of patients who go blind, either temporarily or permanently, is quickly recruited to process tactile information (Cohen, Celnik et al. 1997), particularly when those patients learn quite early to read Braille's embossed characters by rubbing their fingertips on top of them (Sadato, Pascual-Leone et al. 1996). As such, the degree of flexibility and redundancy that these mechanisms confer to a brain entail a level of evolutionary advantage that cannot be rivaled if a central nervous system was defined simply by a digital component.

At this point it is also important to emphasize that the relativistic brain model does not exclude other types of substrate-dependent computations, which could take place at other organizational levels of the brain, for instance, inside an individual neuron, an organelle or a membrane protein.

Moving into the domain of clinical predictions, our theory suggests that most, if not all, neurological and psychiatric disorders could be viewed primarily as disturbances (pathological folding) of the brain space-time continuum. Accordingly, experimental work on animal models of Parkinson's disease (Fuentes, Petersson et al. 2009; Santana, Halje et al. 2014; Yadav, Fuentes et al. 2014) and other neuropsychiatric disorders carried out in MN's laboratory over the past decade suggest that neurological symptoms may emerge as a result of pathological levels of synchronous neuronal firing produced by a subset of brain circuits. In the relativistic framework, that would mean that mental diseases emerge as a consequence of an improper folding of the space-time continuum that forms the "mental space". Accordingly, abnormal levels of neuronal synchrony could disrupt not only the behavior of specific brain circuits, like the motor system in the case of Parkinson's disease, but also impair the generation of optimal neuronal NEMFs that are essential for maintaining other higher order mental functions. This latter effect could account for the profound changes in mood, sense of reality, personality and for symptoms like hallucinations and paranoid thinking that characterize a variety of psychiatric disorders.

In another class of brain disorders, the potential disruption of the proper formation of long-range cortico-cortical or other white matter pathways during early development may impair, among other things, the establishment of the "biological coils" that generate cortical NEMFs. Such a structural failure would irreversibly hinder the process of distributed integration of multi-sensory information into a cortical space-time continuum, a crucial step needed for the formation of a proper "mental space". This disconnection, associated with the development of multiple pools of exacerbated local cortical synchrony, could account for the variety of symptoms observed, for instance, in developmental neurological syndromes. Support for this hypothesis has been obtained in recent brain imaging studies that detect malformation of key cerebral axonal bundles in autistic and cerebral palsy patients (Bakhtiari, Zurcher et al. 2012; Englander, Pizoli et al. 2013).

Temporary disengagement of the same long-range cortico-cortical projections could also explain why one becomes unconscious during the slow wave and spindle stages of sleep (non-REM stage) or during general anesthesia. During slow wave sleep, brain oscillations in the range of 0.1-1Hz dominate the loop formed by the reciprocal connections that link the thalamus and cortex. These oscillations are associated with the stage of drowsiness that precedes the starting of a new sleep cycle. Sleep spindles are characterized by oscillations in the 7-14Hz range and cause a functional disconnection between the thalamus and the cortex, meaning that sensory information gathered in the periphery cannot reach the neocortex. Higher neuronal oscillations, in the so called gamma range (40-60Hz), are correlated with REM sleep, when dreams occur, and full wakefulness. In other words, according to the relativistic brain theory, the perceptual experiences taking place during dreaming and wakefulness require the full engagement of the brain's "biological coils" - i.e. the long-range axonal loops of the white matter – at higher synchronous firing frequency (gamma range) in order to generate the complex combination of NEMFs that, by building the analog computer core of our brains, ultimately accounts for the richness and unpredictability of our conscious experiences. That may also explain why a clear sense of self, the feeling of being, of inhabiting a body that is distinct from the rest of the world, can only be detected in babies, not a birth, but a few months afterwards (Papoušek and Papoušek 1974); that would be the time required for enough of the brain's white matter to mature and become capable of generating the kind of broad NEMFs needed for the sense of self to emerge and manifest itself.

## Brain theories based on electro-magnetic activity

Historically, numerous experiments and theories have highlighted the existence, manifestation and the potential functional roles played by NEMFs. For instant, in a seminal study carried out in 1942, Angelique Arvanitaki showed that, when a giant squid's axons were placed in close proximity in a medium with reduced conductivity, one axon could be depolarized by the activity generated in a neighboring nerve fiber. This classic experiment established, for the first time, the existence of what became known as ephaptic neuronal interactions (Arvanitaki 1942).

More than 50 years later, in 1995, Jeffery investigated the effects on neurons obtained by both endogenous and externally applied electric fields. This author showed that neuronal excitability can be altered at field strengths over a few millivolts per millimeter (Jefferys 1995).

During the 1990s, Wolf Singer's laboratory demonstrated that monkey cortical neurons fired synchronously when two bars displayed on a screen were moved together in the same direction. The same neurons fired asynchronously when the bars were moving in different directions (Kreiter and Singer 1996). A decade later, in 2005, NEMFs were shown to be related to attention and awareness by Andreas Engel, a former member of Singer's group (Debener, Ullsperger et al. 2005).

As findings implicating NEMFs in neuronal firing induction continued to appear in the literature, interest on field-based theories of brain function was rekindled. Thus, in 2000, Susan Pocket published her EMF-based theory of consciousness (Pockett 2000). Almost at the same time, E. Roy John, a New York University neurophysiologist, put forward his own version of a field-based theory of brain function (John 2001). By 2002, Johnjoe McFadden published his first article describing the Conscious Electromagnetic (CEMI) Field Theory (McFadden 2002a; McFadden 2002b). Other theories, such as Fingelkurts' Brain-Mind Operational Architectonics theory (Fingelkurts 2006) and the quantum brain dynamics (QBD) approach of Mari Jibu, Kunio Yasue and of Giuseppe Vitiello, also appeared around this time (Jibu and Yasue 1995).

In McFadden's CEMI, as well as in Fingelkurts' Brain-Mind Operational Architectonics theory, global brain EMF are proposed to influence the electric charges across neural membranes, influencing the firing probability of particular sets of neurons. As such, for some of these authors, of which Susan Pocket and E. Roy John are not included, NEMFs could provide a feedback loop capable of accounting for a variety of higher brain functions, including consciousness and free will (McFadden 2002a; McFadden 2002b).

A brief historical summary of the progression of various EMF-based brain theories would include the following:

1) The **existence** and the detection of EM activity starting in the late XIX century, which led to the first EEG recordings of Berger (Berger, 1929), and up to the 1960s with the first MEG devices.
2) Development of various models of **extracellular fields** starting in the 1960's.
3) Conjecture considering NEMFs as the **basis for consciousness**. Consciousness is here considered as a passive a posteriori recording with no causal efficiency (Pockett 2000).

4) Conjecture of EMF as the mechanism for consciousness and **retroacting by causal efficiency** on the neural network, as proposed by Johnjoe McFadden in the CEMI model (McFadden 2002a; McFadden 2002b).

5) The relativistic brain theory that considers the brain as a space-time continuum (Nicolelis 2011).

Although at first glance our relativistic brain theory resembles McFadden's CEMI, the two theories differ on several key points. Among those, we could cite the concept of the "brain's own point of view", the genesis of "anticipatory expectations" rather than a posteriori binding, the key proposition that distinct analog computers emerge from the brain at any given moment in time, and that EFMs generate a space-time continuum from which the mental space emerges. Despite these and other differences, the relativistic brain theory does support most of the conclusions put forward by the CEMI model. Moreover, all the experimental confirmations cited in McFadden's articles apply to the relativistic brain theory as well.

It important to close this section by stating that many authors, such as William R. Uttal, Jeffrey Gray and Bernard Baars, have criticized EM field theories (Gray 2004; Uttal 2005). Our reply to these criticisms will be presented elsewhere.

Finally, additional support for the role of NEMFs in higher order brain processing comes from experiments carried out with brain-machine interfaces (BMIs) (see Chapter 1) in primates. In this paradigm, monkeys learn to use electrical brain activity, recorded from a random sample of 100-500 cortical neurons, to directly control the movements of artificial actuators, like robotic or virtual arms/legs. Sensory feedback originated by the movement of these actuators can then be transmitted back to the subject through either visual or tactile signals or even direct electrical micro-stimulation of the animal's cortex (O'Doherty, Lebedev et al. 2011). Monkeys quickly learn to use BMIs to control the movements of such actuators without producing overt movements of their own bodies.

Figure 2.4 illustrates an interesting phenomenon observed during the initial phases of these BMI experiments when the animals are learning to use their brain activity alone to move the actuator. Inspection of the correlation matrix depicted in this figure reveals that, during this initial training phase, there is a significant increase in neuronal firing correlation among the subset of cortical cells selected to feed the computational algorithm employed to move the artificial actuator.

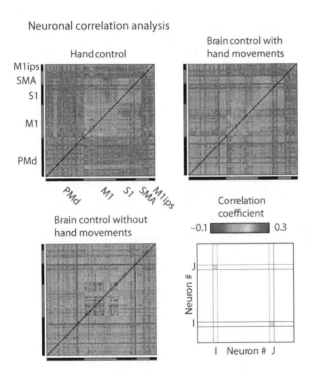

*Figure 2.4 – Analysis of pairwise correlations in firing between populations of neurons recorded simultaneously in the same rhesus monkey using a brain-machine interface to control a robotic arm. Correlations increased significantly when animals move from hand control of a joystick (to control a computer cursor), to the use of their own brain activity alone (brain control mode) to perform the same. Two types of brain control mode were tested: one in which some hand movements were allowed and another in which no hand movements were present. Neuronal correlations increased significantly from the former to the latter mode of brain control. The highest correlations were*

*between the neurons recorded in the same cortical area. Mlips, primary motor cortex, hemisphere ipsilateral to the working hand; PMd, dorsal premotor cortex; sl, primary somatosensory cortex; SMA, supplementary motor area. All data derived from Carmena, Lebedev et al. 2003. Published with permission from Nature Publishing, originally appeared in Nicolelis MAL, Lebedev MA. Principles of Neural Ensemble Physiology Underlying the Operation of Brain-Machine Interfaces. Nat. Rev. Neurosci. 10: 530-540, 2009.*

This increase in neuronal firing correlation occurs both within and between cortical areas from which the individual neurons are randomly sampled. As the animal becomes proficient in operating a BMI, the levels of neuronal firing correlation decreases. This transient increase in neuronal firing correlation is consistent with the global action of a new "analog computation engine", generated by NEMFs produced by the brain. According to the relativistic brain theory, this analog engine would underlie multiple functions, such as "a priori binding" the subset of cortical neurons sampled by the experimenter into a true "functional unit", which is in charge of generating the motor signals needed to control the actuator. The analog computational engine would also provide a way through which the artificial actuator (e.g. robotic arm or legs, virtual avatar) can be assimilated, as a true extension of the subject's internal body schema, through the updating of body-related memories stored throughout the digital component of the cortex (e.g. neuronal networks).

Figure 2.5 illustrates another curious finding derived from the same BMI experiments. These observations come from the analysis of neuron-dropping graphs illustrated in Figure 1.1. These graphs relate how much variance of a given motor parameter, such as hand position or gripping force, can be accounted for by a linear model, which is used to combine the firing rates of a random sample of cortical neurons recorded simultaneously. Interestingly, we have observed that no matter which cortical area our neuronal samples come from, the shape of these neuronal dropping curves always tends to follow a logarithmic function. Given a certain motor parameter, the only thing that changes when one compares different cortical areas is the slope of the logarithm curve. In other words, the estimation of a kinematic parameter, using linear models, seems to improve with the log of the total number of cortical neurons recorded simultaneously.

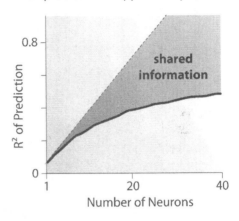

*Figure 2.5 – A thick line depicts a neuronal dropping curve obtained from a sample of 40 simultaneously recorded neurons in the primary somatosensory cortex of a rhesus monkey. The curve depicts the relationship between number of neurons and the prediction accuracy (measure by R2) of upper limb position during use of a brain-machine interface. As reported in many studies, this relationship is defined by a logarithm curve. Dashed line represents a theoretical straight line representing the same relationship (number of neurons vs prediction capability). The shade area indicates the amount of shared information between the simultaneously recorded neurons. One of the hypothesis to explain why the experimental curve (thick line) does not follow a straight line is that cortical neurons share a significant amount of "biological noise". The RBT proposes that part of this noise is generated by NEMFs.*

The red shaded area in Figure 2.5 indicates how much each of these logarithmic curves diverges from a straight line. The usual explanation for this divergence is that since cortical neurons share some "correlated noise", as one increases the number of individual neurons added to our linear models the resulting curves asymptote, approximating a logarithmic function. Until now the potential source of this "correlated noise" remained unknown. Here, we propose that at least a fraction of this shared variance arises from the concurrent action of cortical NEMFs acting throughout the entire neocortex.

Having introduced the relativistic brain theory and briefly reviewed its main tenets, we now turn to the discussion of why we think this new view of the human central nervous system invalidates any attempt to simulate complex animal brains using a Turing machine.

## CHAPTER 3 – The mismatch between integrated systems like brains and Turing machines

Over the past 6 decades, the pervasive application of digital computing in all aspects of our lives has generated the common belief, among scientists and society alike, that every single natural physical phenomenon can be reduced to an algorithm that runs on a digital computer, the most popular application inspired on the concept of the universal Turing Machine, proposed by the British mathematician and logician Alan Turing (Turing 1936). This hypothesis, largely borrows its credibility from the classic Church-Turing assumption that "any function which would be naturally considered as "computable" could be computed by a universal Turing machine". Essentially, the key source of confusion in this latter statement resides in the definition of the word "naturally"[4].

Although the Church-Turing hypothesis only concerns mathematical modeling (formal systems), many authors interpreted it as if it would set a computation limit for all natural phenomena, meaning that no physical computing device could exceed the capacity of a Turing Machine. Forgotten in this assumption is the fact that Turing's computability relates to questions arising in formal mathematics. As such, Turing's computation theory makes numerous assumptions that restrain its applicability to biological systems. For example, it assumes that the representation of information is formal, i.e. abstract and syntactic, rather than physical and semantic as is the case for most biological systems. This assumption, therefore, gives the erroneous impression that bits and bytes of information can

---

[4] The Church Turing thesis concerns a method in logic and mathematics. If M is such a method it should be effective or mechanical meaning that M is expressed in a finite number of precisely given instructions, themselves formulated in a finite number of symbols. To be effective, M should also produce a result in a finite number of steps that can be carried out by a human being with no insight, intuition or other ability than applying M.

accurately represent the broad range and scope of the mental processes that emerge from the brains of animals[5].

In recent years, a large group of computer scientists and neuroscientists has adopted the "physical version" of the Church-Turing assumption as the main theoretical basis to propose that any animal brain, including our own, can be reduced to an algorithm and simulated on a digital computer. Accordingly, the successful approach of using simulations to the study of mechanical systems has been consequently extended to the study of biological systems whose complexity is far superior. This philosophical attitude is known as computationalism[6] and has been defended by many philosophers such as Jerry Fodor (Fodor 1975) and Hillary Putnam (Putnam 1979)[7]. Yet, critics of computationalism regard as "purely mystical" the view that high-level brain functions, involving language-understanding, decision making, and reasoning will somehow "emerge from the interaction of basic behaviors like obstacle avoidance, gaze control and object manipulation" (Copeland 2002; Copeland, Posy et al. 2013). At the limit, computationalism not only predicts that the entire spectrum of human experiences can be reproduced and initiated by a digital simulation, but it also implies that, in a near future, because of exponentially growing computer power, machines could supplant the totality of human mental capabilities. This latter notion, put forward by Ray Kurzweil and others, has become known as the Singularity hypothesis[8] (Kurzweil 2005). Overall, this hypothesis not only

---

[5] Further in depth discussion on the misunderstandings of the Church-Turing thesis is available at the Stanford Encyclopedia of Philosophy (http://plato.stanford.edu/entries/church-turing)

[6] This name is attributed to Hilary Putnam In « Brains and behaviors » 1961

[7] For a broader discussion on computationalism see : Gualtiero Piccinini : » Computationalism in the philosophy of mind » Philosophy compass 4 (2009).

[8] In his book The Age of Spiritual Machines: When Computers Exceed Human Intelligence, Kurzweil states a radical version of Church-Turing: "If a problem is not solvable by a Turing Machine, it is also not solvable by a human mind.

provided some relief to research programs, including the one known as strong Artificial Intelligence, which has come significantly short of fulfilling previous optimistic predictions[9], but it has also served as the basis for a series of recent proposals to simulate entire brains using supercomputers[10].

While we certainly do not doubt that brains and other organisms process information, a series of arguments listed below refute the notion that such processing can be reduced to algorithms and be meaningfully simulated on a digital computer, or any other Turing Machine for that matter.

Many neuroscientists today believe that higher neurological functions in both animals and humans derive from complex emergent properties of the brain, even though the origin and nature of these properties remain debatable. Emergent properties are usually considered as global system attributes, which do not result from the description of its individual components. Such emergent properties occur everywhere in nature where elements interact and coalesce to form an entity – like a flock of birds, a school of fish or a stock market – normally designed as a complex system. Hence, the investigation of complex systems has become the focus of a large spectrum of disciplines, from natural science (e.g. chemistry and biology) to the humanities (e.g. economy and sociology) (Mitchell 2009).

The central nervous system (CNS) of animals can be considered as a typical example of a complex system. CNS complexity, however, extends to its different organizational levels, i.e. from its molecular, cellular and circuitry scaffolding, all the way to the entire nervous systems as a whole. Thus, to be really precise in our modeling of a particular animal's CNS, we should also include in the definition of its complexity description its exchanges with external entities, such as the surrounding

---

[9] In 1968 Marvin Minsky, head of MIT AI lab had announced : « Within a generation we will have intelligent computers like HAL in the film 2001. » Certainly, his prediction did not materialized and Minsky has recently declared that brain simulation programs have very little chance to succeed.

[10] Like the IBM Brain Project and the EU Human Brain Project.

environment as well as other subject's brains, since these also interact and continuously modify the particular CNS under investigation.

Brains also exhibit another very important property, the ability to constantly reorganize themselves – at both the morphological and functional levels (see Chapter 1) – as a result of current and past experiences. In other words, information processed by the brain is used to reconfigure its structure and function[11], creating a perpetual recursive integration between information and brain matter. That is the reason we usually refer to systems like the CNS as complex adaptive systems.

Importantly, as we shall examine hereunder, the very characteristics that define a complex adaptive system are the ones that undermine our capacity to accurately predict or simulate its dynamic behavior. For example, at the beginning of last century, Poincaré showed that the emerging behaviors of a system composed of even a few interconnected elements – let alone tens of billions of hyper connected neurons - cannot be formally predicted through the analysis of their forming elements (Poincaré 1902). In a complex system like the brain, individual elements (i.e. neurons) dynamically interact with one another in order to generate new behaviors of the system as a whole. In return, such emergent behaviors directly influence the system's various elements, pretty much like an orchestra whose individual instruments (elements) are continuously reshaped by the very music (emergent property) they play.

Here, we will argue that the brain has to be viewed as an *integrated* system, a particular "continuum" that processes information as a whole and for which one can distinguish neither "software" from "hardware", nor "memory" from "processing"[12]. Instead, in our view, the way a brain represents information is deeply related to the functioning of its various levels of physical organization from its global macrostructure down to its quantum level. As such, the way animal brains generate, represent,

---

[11] This capacity of information to act on matter is often called causal efficiency.

[12] Some authors use the expression « embedded information ».

memorize and handle information is significantly different from the way computer scientists normally conceptualize how various material realizations of the universal Turing Machine, such as digital computers, handle computing through the employment of algorithmic programs (software) dissociated from the machine's hardware.

In this new context, when one examines the brain's operations by using both a mathematical and a computational point of view, emergent behaviors cannot be fully rendered via classical, syntactically abstracted software procedures, running on fixed hardware. In other words, the rich dynamic semantics that characterize brain functions cannot be reduced to the limited algorithmic syntax employed by digital computers. That happens because emergent properties that simultaneously encompass different levels of the brain's physical organization, involving the precise coordination of billions of top down and bottom up interacting events, are not effectively computable. Instead, they can only be temporarily approximated by a digital simulation. Thus, if one accepts that brains behave like *integrated* and adaptive complex systems, these approximations will immediately diverge from the behavior of the system and the emergent properties that they intended to simulate, making it impossible to reproduce the brain's main functional attributes. That means that the typical strategy used by modelers will neither describe nor reproduce the full complex dynamic richness that endows living brains, including our own, with their ultimate functions and capabilities.

As it will become apparent in the next chapters, our central thesis is also supported by the argument that models running on Turing machines cannot grasp the complexity of higher CNS functions, simply because they cannot simulate the type of analog integrated computations that generate these functions in real brains, where everything is simultaneously affecting everything. What follows, therefore, is an exposition of the mathematical, computational, evolutionary, and neurobiological arguments that support the contention that no animal brain, complex enough to merit scientific investigation, can be reduced to a simulation on a Turing machine[13].

---

[13] In this monograph we shall use the word computation to designate any form of information processing, not necessarily limited to Turing computation.

# CHAPTER 4 – The mathematical and computational arguments against the possibility of simulating brains in a Turing machine

Before describing the mathematical and computational objections that form our central argument, it is important to spell out what exactly defines a computer simulation of a natural system, like a brain (Figure 4.1), on a digital computer.

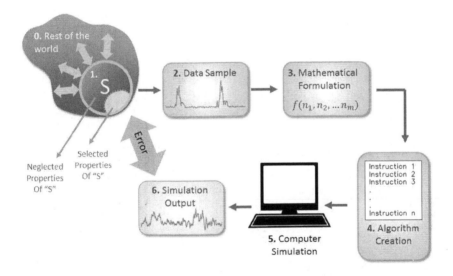

*Figure 4.1 – Multiple steps involved in the process of simulating the behavior of a biological system "S" using a digital computer. First, it is important to realize that "S" is in constant interactions with the rest of the world (as represented by bidirectional arrows linking S to it). To start the process, some properties of "S" have to be selected for the simulation (1). This, however, leads to the neglect of a large set of other properties of "S". Next, some data – usually, the variation of the magnitude of a parameter in time - have to be sampled from "S" to serve the basis for the simulation (2). Now, it is the time to select a particular mathematical formulation that can describe the behavior of the "S" parameter sampled (3). Once the mathematical formulation is selected, one needs to reduce it into an algorithm that can be fed into a computer (4). Now, the computer will run the algorithm (5) and generate a simulation output (6). Notice that there is an error between this simulation*

*output and the data sample and an even greater discrepancy when the simulation output is compared to the behavior of "S".*

Throughout the process of building a digital simulation, many preconceptions and assumptions (0) are made (such as the type of information representation) and various obstacles have to be overcome. These assumptions may at the end completely invalidate the model. For example, let us consider any physical system S whose evolution we want to simulate. The first approximation is to consider S as an isolated system. Yet, biological systems cannot be isolated without losing many of their functionalities. For instance, if S is a living system (1), also known as a dissipative system according to Prigogine (Prigogine 1996), its structure at any given moment is totally dependent on its exchange of matter and information with its environment. In this case, S processes information as an integrated system (see below). Considering S as an isolated system, therefore, can completely bias the simulation, especially when a living organism such as the brain is considered. That constraint would, for instance, invalidate any attempt to build a realistic model of a living, adult animal brain based on data collected (2) from experimental preparations, like in-vitro brain slices obtained from juvenile animals. Since this experimental preparation dramatically reduces the true complexity of the original system (the living brain), as well as its interactions with the surrounding environment, translating results obtained from such a reduced model to the real behavior of a living brain is simply meaningless, even when the model yields some trivial emergent behavior, like neuronal oscillations (Reimann, Anastassiou et al. 2013).

The next step in a computer simulation involves selecting the data measured from S and their type, knowing that we are neglecting a wide variety of other data and computations at different observation levels that we, by option or necessity, consider irrelevant. Yet, in case S is an *integrated* system, like a brain, one can never be certain that some further observation levels are truly irrelevant or require, for instance, the introduction of a quantum description.

Once observations or measurements are made about the behavior of a given natural phenomenon related to S, one then tries to select a mathematical formulation (3) that can fit the selected data. As a rule, this mathematical formulation is defined by a set of time dependent differential equations[14]. However, it is important to emphasize that, in most cases, this mathematical formulation is already an approximation, which does not render the natural system completely at all its many organizational levels. As expressed by Giuseppe Longo (Bailly and Longo 2011): most physical processes simply cannot be defined by a mathematical function.

The final step involves trying to reduce the chosen mathematical formulation to an algorithm (4), i.e. a series of sequential instructions that can run on a digital machine. Altogether, that means that a computer simulation is an attempt to simulate the mathematical formulation of a set of observations made of a natural phenomenon, not the natural phenomenon itself. Since the evolution of a biological system is not governed by the binary logic used in a digital computer (5), the outcome obtained (6) by a computer simulation may, in many circumstances, evolve very differently than the natural phenomenon itself. This is particularly true when one considers complex adaptive systems where emergent properties are essential for the proper operation of the whole system. Given that we have no evidence that a system governed by binary logic (5) will produce the same kind of emergent properties as the original phenomenon, at the end of the process of digital simulation, one may discover that, at best, one's algorithmic solution may only provide an approximation to the mathematical formulation and, hence, to the natural phenomenon. But since this approximation could diverge quickly from the real behavior of the natural system, the digital simulation may yield only meaningless results from its very beginning. Indeed, most models that claim to have created "artificial life" employ combinations of various

---

[14] One should notice that these mathematical structures have generally been developed for applications in physics. As such, they not necessarily apply well to biological systems.

algorithmic techniques, from object oriented, to process-driven, to iterative grammars, in order to try to mimic human behavior. According to Peter J. Bentley (Bentley 2009), this is a flawed strategy, since:

> *"there is no coherent method to correlate these programmer tricks with biological entities. As such, this approach results in opaque and largely unsustainable models that rely on subjective metaphors and wishful thinking to provide relevance to biology..."*

The mathematician Michael Berry has proposed a simple example to illustrate the difficulties related to simulating any physical system. The example is based on trying to simulate the successive impacts of a billiard ball in a pool game, knowing the complete initial conditions. Calculating what happens during the first impact of the billiard ball is relatively simple if you have collected the parameters and estimated the strength of the hit. However, estimating the second impact gets more complicated since one has to be more precise in estimating the initial states in order to get an acceptable estimation of the ball's trajectory. To compute the ninth impact with great precision, you will need to take into account the gravitational pull of somebody standing near the table. And to compute the 56th impact, every single particle in the universe needs to be present in your assumptions; an electron on the other side of the universe, must figure into your calculation. The difficulty is even higher with living systems. Thus, in the case of a brain that requires exquisite coherence from billions of neurons and multiple levels of organization to exert its functions, the possibility of one's simulation diverging is overwhelmingly high.

**The mathematical argument: computability**

Computability is specifically related to the Turing Machine's model of computation since it refers to the possibility or not of translating a mathematical formulation to an effective algorithm. Computability is thus an alpha-numerical construction and not a physical property. Since most mathematical formulations of natural phenomena cannot be reduced to an algorithm, they are defined as non-computable functions. For instance, there is no general procedure that allows a systematic debugging of a digital computer. If one defines a function F which would examine any given program running on a given machine and which would take the value 1 each time it finds a bug and zero otherwise, F would be non-computable. Non-computability here is illustrated by the fact that there is no algorithmic expression of F that can detect in advance any possible future bug that may hamper the work of a computer. Whatever one does, the machine will always exhibit unexpected faulty behaviors that could not be predicted when the computer and the software were manufactured.

It is also well know that there is no such a thing as universal anti-virus software[15]. The reason for this is because the function F whose output is all programs that do not contain a virus is also non-computable. The same type of reasoning also justifies why there is neither a universal encryption system on a digital machine, nor algorithmic procedures to tell which dynamical systems are chaotic or not[16].

Now, we can cite our first argument against the possibility of simulating an animal brain, or some of its complex functions, in a digital computer. Its formulation goes like this:

*"A living animal brain can generate some behaviors that are only fully described by non-computable functions. Since those cannot be dealt with properly by a Turing Machine, there is no possibility of simulating precisely a brain on a digital computer, no matter how sophisticated it is."*

---

[15] This is a corollary of Rice's theorem : Any non trivial property about a language recognized by a Turing Machine is undecidable. (Rice, 1953 : Classes of recursively enumerable sets and their decision problems.)
[16] Nor is there any method to determine whether a Diophantine equation has solutions or not (Yuri Matayasevich, 1970).

The examples above represent just a minute sample of the pervasiveness of non-computability in mathematical representations of natural phenomena. These examples are all consequences or variants of the famous Hilbert's halting problem (see below). In his classic 1936 article (Turing 1936), Alan Turing demonstrated that an algorithmic machine, which is now known as the Turing Machine and served as the theoretical prototype of modern digital computers, cannot solve the halting problem. Accordingly, Hilbert's halting problem has since become the primordial model of non-computable functions.

In reality, most functions are non-computable because there are numerable possible Turing machines and an infinitely larger non-numerable number of functions. This happens because it is possible to enumerate the set of all program codes, i.e. there are only a numerable number of programs and, consequently, a numerable number of Turing Machines.

The halting problem expresses that there is no way of deciding in advance which functions are computable and which are not. This is also why the Church-Turing hypothesis remains a hypothesis, i.e. it could never be proved or disproved by any Turing Machine. Actually, nearly all functions cannot be computed by a Turing machine, including the majority of functions that should be used to describe the natural world and, in our view, those generated by highly evolved brains.

Being already aware of the limitations of his Turing Machine, in his PhD thesis published in 1939 (Turing 1939), Alan Turing himself attempted to overcome them by conceiving what he called an Oracle Machine. The whole point of the Oracle Machine was to introduce a real world tool for reacting to what "*could not* be done mechanically" by the Turing Machine. Thus, the Oracle Machine provided an "external advisor" that would be consulted and activated at some steps in the calculation that could not be solved algorithmically by the Turing Machine. After the Oracle's answer, the computation resumed. In another words, the Oracle should be called to solve a non-computable step.

Interestingly, Turing himself suggested that these Oracle machines could not be *purely mechanical*. Indeed, the whole point of the Oracle Machine was to explore the realm of what *cannot* be done by purely algorithmic processes. Accordingly, Turing showed that some Oracle Machines are strictly more powerful than Turing machines. He went on and stated that:

*"We shall not go any further into the nature of this Oracle apart from saying that it cannot be a machine."*

Turing seminal work, therefore, launched the field of hypercomputing. Turing himself never suggested that such computers could be built, but he repeatedly insisted that 'intuition' (a non-computable human property) is present in every part of a mathematician's thinking. For instance, when a mathematical proof is formalized, intuition has an explicit manifestation in those steps where the mathematician sees the truth of a formerly unprovable statement. Turing, however, did not offer any suggestion as to what, in his opinion, the brain was physically doing in a moment of such 'intuition'.

Many decades after the introduction of the Oracle Machine, Chaitin, da Costa and Doria (Chaitin, da Costa et al. 2011), as well as other authors (Pour-El and Richards 1989), proposed the idea that *"analog devices, not digital ones, can decide some undecidable arithmetic sentences"*. That would happen because analog computational engines "physically compute", meaning they "compute by simply obeying the laws of physics, rather than by running a pre-given algorithm within a formal system, created to solve equations that are meant to describe a system's behavior. Put differently, in analog computers there is no separation between hardware and software, because the computer's hardware configuration is in charge of performing all the computing and can modify itself. This is precisely what we have defined above as an integrated system.

According to Chaitin, da Costa and Doria, analog devices could serve as the basis of hypercomputers, i.e. "real-world devices that settle questions which cannot be solved by a Turing Machine". These authors further suggest that the possibility of effectively building a prototype of such a hypercomputer, by coupling a Turing machine with an analog device, is only a matter of developing the proper technology, reducing the question to an engineering problem. As such, both Turing's work and the theoretical possibility of building hypercomputers raise the notion that there are physical objects having capacities greater than the Turing Machine. Therefore, it is not surprising at all that integrated systems, such as the brain, do overcome the computational limitations of the Turing Machine. Indeed, the very existence of animal brains can be used to disprove the "physical version" of the Church-Turing hypothesis[17].

Overall, these arguments further boost our thesis that animal brains cannot be effectively simulated on digital computers. In fact, Turing himself insisted repeatedly on the limitations of his machine:

*"The idea behind digital computers may be explained by saying that these machines are intended to carry out any operations which could be done by a human computer.* (Turing 1950)" or again *"The class of problems capable of solution by the machine [the ACE] can be defined fairly specifically. They are [a subset of] those problems, which can be solved by human clerical labour, working to fixed rules, and without understanding* (Turing 1946)".

In his book "The fabric of reality," the physicist David Deutsch argues for a very strong physical version of the Church-Turing thesis (Deutsch 1997). The ultimate reality being described by the laws of quantum physics, it should be possible to simulate on a universal quantum computer the totality of all natural phenomena. According to this view, simulating a brain on a quantum or digital computer should theoretically be possible.

Recently, David Deutsch has softened his position by recognizing that the brain has capacities exceeding those of both

---

[17] For instance, in his book The fabric of reality, (1997), David Deutsch has argued that there are virtual reality environments that Turing machines cannot simulate by using a diagonalisation argument.

digital and quantum computers (see our Gödelian argument hereunder). According to him:[18]:

*"It is uncontroversial that the human brain has capabilities that are, in some respects, far superior to those of all other known objects in the cosmos".*

## Complexity and non integrable functions

According to Poincaré (Poincaré 1905), complex dynamical systems (in which the individual elements are themselves complex interacting entities), cannot be described with integrable functions, i.e. derivative functions that can be integrated allowing us to figure out relations between the quantities themselves. Such a dynamical system is characterized in terms of the sum of the kinetic energy of its particles, to which one has to add the potential energy resulting from the particles' (elements) interactions. In fact, this second term is responsible for the loss of linearity and integrability of these functions. Poincaré not only demonstrated the non-integrability of these functions, but he provided an explanation for it: the resonances between the degrees of freedom. That means that the richness of those complex systems' dynamic behaviors cannot be captured by solvable sets of simple differential equations, because their resonances will in most cases lead to the appearance of infinite terms.

Brains are formed by intrinsically complex, self-adaptable (i.e. plastic) elements, whose elaborate connectivity and integration adds many other levels of complexity to the entirety of a nervous system. Furthermore, the behavior of each neuron, at the various observational level of the embedding neuronal ensemble, cannot be understood except with reference to the global pattern of brain activity. As such, even the most rudimentary animal brain fulfills Poincare's criteria to be considered a complex dynamical system with resonances

---

18 Aeon Magazine (Oct 2012).

between different organization levels or composing biological elements (e.g. neurons, glia, etc.) for which there is no integrable mathematical description.

Moreover, if one assumes that vital computations in the brain, indeed the ones which are responsible for its emergent properties, are taking place, even partially, in the analog domain, it follows that a digitalization process would be capable of neither approximating the brain's physiological behavior at a precise moment in time, nor predicting how it would evolve in the immediate future.

Poincaré also showed that dynamical complex systems can be very sensitive to initial conditions and are subject to instabilities and unpredictable behaviors, a phenomenon known today as chaos (Poincaré 1905). In other words, to make a prediction with a digital machine, concerning the behavior of a Poincare's time-varying analog system, one would need to know precisely the initial state of the system and have an integrable computable function that can compute a prediction of its future state. Neither of these conditions can be met when we talk about brains. Given the inherently dynamic nature of nervous systems, it is impossible to estimate precisely the initial conditions of billions of neurons, at various organizational levels; every time a measurement is taken, the initial conditions change. Moreover, most of the equations selected to describe the brain's dynamic behavior would be non-integrable.

In light of those constraints, typical simulations on a Turing Machine, even if that machine is a modern supercomputer with thousands of microprocessors, are not likely to reveal any relevant physiological attributes of real brains. Essentially, such simulations will likely diverge from the dynamic behavior of real brains as soon as they start, rendering their results absolutely useless for the goal of learning something new about how brains operate.

## The computational argument: tractability

Simulating the brain on digital machines also involves dealing with numerous non-tractable problems. Tractability in a digital computation relates to the number of computer cycles required to conclude a given calculation as well as other physical limitations, such as available memory or energy resources. Thus, even if an algorithmic representation of a mathematical function that describes a natural phenomenon can be found, the computing time required to run a simulation with this algorithm may not be viable in practical terms, i.e. it may require more than the life of the entire universe to yield a solution. These kinds of problems are known as non-tractable. Since a Universal Turing Machine, such as a digital computer, is capable of solving any problem that another Turing Machine could solve, the simple increase in computing power or speed does not transform a non-tractable problem into a tractable one. It can only make a better approximation in a given time.

Let's examine an example of a non-tractable problem. Protein structures embedded in the neuron's membrane, known as ion channels, are fundamental for the transmission of information between brain cells. To enact their effects proteins have to assume a particular optimal tridimensional configuration (Anfinsen 1973). The final 3D shape of proteins, achieved by a process known as protein folding, which is a critical element for the proper function of neurons, includes expanding, bending, twisting, and flexing of the amino acid chain that forms the protein primary structure. Each individual neuron has the potential to express some 20,000 different protein coding genes as well as tens of thousands of non-coding RNA. As such, proteins are part of the integrated system that generates information in brains. Let us then consider a simple protein formed by a linear sequence of about 100 amino acids and suppose that each amino acid can assume only 3 different conformations. According to the minimum energy model, normally used to attempt to estimate the three dimensional structure of proteins, we would have to examine $3^{100}$ or $10^{47}$ possible states to reach a final result. Since the solution landscape of our protein folding model grows exponentially with the number of amino acids and with the number of considered conformations, this becomes a non-tractable problem; if the protein should find its native state by random search, visiting one state each picosecond, the overall search could take longer than the age of the universe.

This example illustrates well what Turing intended by a "real world Oracle": in real life, a protein, the integrated system, solves the problem in milliseconds, whereas the algorithmic computer translation can take more time than the whole life of the universe. The difference here is that the "protein hardware" computes the optimal solution and "finds" its 3D configuration by simply following the laws of physics in the analog domain, while a Turing Machine would have to run an algorithm created to solve the same problem on a digital device. As we will examine in the section dedicated to evolutionary arguments, real world organisms, being integrated systems, can handle their complexity in an analog way that cannot be captured by a formal system, ergo neither by algorithms.

Protein folding is an optimization problem, i.e. it involves searching for an optimal solution in a landscape of possible solutions. This is usually expressed as the minimum or maximum of a mathematical function. Most optimization problems happen to fall in the category of intractable problems, usually named NP hard problems[19]. All the problems that complex brains are good at solving fall in this category. In simulations, these problems are generally dealt with by using approximation algorithms that could give near optimal solutions. In the case of a brain simulation, however, an approximation solution would have to be found simultaneously at different organizational levels (e.g. molecular, pharmacological, cellular, circuit, atomic quantum level, etc.), making the question even more complicated since optimizing a complex adaptive system often implies suboptimizing its own subsystems. For instance, by limiting the organizational levels that one considers in simulating an integrated system, as is traditionally done during a coarse grained

---

[19] NP problems are those problems for which solutions can be checked in polynomial time by a deterministic Turing machine. NP hard problems are a class of problems that are "at least as hard as the hardest problems in NP".

simulation of a brain, one would likely miss crucial phenomena, lying at lower levels of the integrated system, which can be critical for the whole system optimization.

Usually, tractable algorithms are designed as approximations in order to allow some estimation of future states of a natural system, given some initial conditions. This is, for instance, how meteorologists attempt to model the weather and make predictions whose probabilities of realization are known to rapidly decrease with time. In the case of simulations of the brain, the tractability problem becomes even more critical because of the huge number of interconnected elements (neurons) interacting in a precise time sequence. For instance, given that a digital computer has a clock, which runs on a step by step function, the problem of updating, in a precise timely order, billions or even trillions of parameters, which define the current state of the brain, becomes totally non-tractable. Yet again, any further attempt at predicting the next state of a brain, from arbitrarily chosen initial conditions, will produce a poor approximation. Consequently, no meaningful predictions of the emerging properties can be obtained in the long run, even in time-scales as short as a few milliseconds.

Moreover, if one accepts the notion that there is some fundamental aspect of brain function mediated by analog fields (see below), a digital machine would neither be able to simulate these functions, nor be capable of updating all the huge parameter space (billions or trillions of operations) in precise synchrony during the same clock cycle. In other words, a digital simulation would not generate any realistic brain emergent property.

At this juncture, we should observe that if one wishes to simulate a whole brain, i.e. a dissipative, highly connected system interacting with the animal's body and the external environment, any processing speed that does not exactly match in real time should be automatically disqualified. A brain simulation running at a speed - even at a supercomputer speed – that is lower than the "real" environment to which it is connected, and with which it will be in constant interaction, will not produce anything similar to what a naturally evolved brain can produce or, for that matter, feel. For instance, a real animal brain must detect, in a fraction of a second, whether it is about to be attacked by a predator. Then, if one's "simulated brain" reacts at a much slower speed, the simulation will not be of any practical use to understand how brains deal with the natural phenomenon of predator-prey interaction. These observations apply to a broad spectrum of CNSs in the phylogenetic scale, ranging from the most rudimentary brains of invertebrate animals, such as C. elegans, which only contains a few hundred neurons, all the way to our own brains, which are formed by up to 100 billion neurons.

## CHAPTER 5 – The evolutionary argument

To some degree the mathematical and computational arguments laid out in the previous chapter should not be surprising as they rely on Gödel's and Turing's work in the 1930's. Gödel himself sustained the idea that his incompleteness theorems provided a precise and explicit indication that the human mind exceeds the limitations of Turing Machines and that algorithmic procedures could not describe the entirety of the human brain's capacities:

*"My theorems only show that the mechanization of mathematics, i.e. the elimination of mind and abstract entities, is impossible if one wishes to establish a clear foundation. I have not shown that there are non-decidable questions for the human mind, but only that there are no machines that can decide all questions of number theory" (letter from Gödel to Leon Rappaport 2 August 1962).*

In his famous Gibbs lecture[20], Gödel also asserted the belief that his incompleteness theorems imply that the human mind far exceeds the power of a Turing Machine.

Copeland, Posy and Shagrir (Copeland, Posy et al. 2013) also argue that Turing proposed what he called the multi-machine theory of mind. According to this theory, mental processes form a finite procedure that is not necessarily mechanical. Lucas (Lucas 1961) and Penrose (Penrose 1991) further commented on what is now known as the Gödelian argument. It goes like this:

**Gödel's incompleteness theorems (Gödel 1931) can be interpreted as giving a clear indication of the limitations of a formal system that do not affect the human brain, since the central nervous system can generate and establish truths that**

---

[20] Some basic theorems on the foundations of mathematics and their implications, Collected work III, Oxford University Press, 1951/1995 pp 304-323.

**cannot be proven to be true by a coherent formal system, i.e. an algorithm running on a Turing Machine.**

Roger Penrose has expressed the first incompleteness theorem in a way that clearly relates it to a specific, non-computable capacity of the human brain: belief.

*"If you **believe** that a given formal system is non contradictory, you must also **believe** that there are true proposals within the system that cannot be proved to be true by the formal system".*

Penrose has maintained that the Gödelian arguments offer a clear indication of some kind of limitation to digital computers that is not otherwise imposed on the human mind. In support of Penrose's position, Selmer Bringsjord and Konstantine Arkoudas (Bringsjord and Arkoudas 2004) have given very convincing arguments to sustain the Gödelian thesis, by showing that it is possible that human minds work as "hypercomputers". This means that the human brain can exhibit capacities – like recognizing or believing that some statement is truthful - that cannot be simulated by an algorithm running on a Turing Machine. That would imply that the full richness of human thoughts cannot be reduced to systems running algorithms. As far as we can tell, these arguments have not been refuted yet.

Accordingly, the central premise of the Singularity's hypothesis, i.e. that a future generation of Turing Machines will outperform the mental capacity of human brain, can be falsified at once simply by saying that no digital machine will ever solve Gödel's riddle.

**The evolutionary argument**

Proposals to reverse-engineer or simulate animal brains in Turing Machines also ignore a fundamental difference between an organism and a mechanism, such as a digital computer. Mechanisms are engineered and intelligently built according to a pre-existing plan or a blueprint. That is why a mechanism can be encoded by an algorithm and, consequently, be reversed engineered. Organisms, on the other hand, emerge as the result of a huge number of evolutionary steps, happening at multiple levels of organization (from molecules to the entire organism), which do not obey any previously established plan. Rather, these steps take place through a series of random events. Organisms, therefore, are closely related to their environment. Indeed, Prigogine, in his book 'The end of certainty' (Prigogine 1996), calls them dissipative systems, because their material organization, at any given moment, is totally dependent on energetic, material and informational exchanges with their surroundings.

Organism can only exist far from thermo dynamical equilibrium. Thus, the information they generate about themselves and the surrounding world has to be used to constantly maintain a local negative entropy (Schrödinger 1944)[21]. That task can only be achieved by continuously reshaping and optimizing the organism's matter substrate from which this information has emerged. Otherwise, without the perpetual expression of this "causal efficiency"[22], the organism would progressively disaggregate. This is obviously true for the brain. Thus, the idea of 'substrate independent information'[23] (at the root of the reverse engineering concept) cannot apply when considering the information flow within organisms. Not only is information flow in organisms, and especially in the brain, handled at a large series of different organizational levels, but it continuously modifies the material subtract (e.g. neurons, dendrites, spines, proteins,) that has generated it. This unique process binds *both organic matter and information in an irreducible single entity. Information in organisms is substrate-dependent, a conclusion that confirms the integrated nature of*

---

[21] For Schrödinger living matter evades the decay to thermodynamical equilibrium by homeostatically maintaining negative entropy in an open system. Today, we call this information.
[22] Causal action of information on matter.
[23] Information considered as independent of the matter that encodes this information, such as software in a typical digital computer.

*the brain and overtly exposes the unsurpassable difficulties of applying the "software/hardware dichotomy" to surmise the way in which an animal's CNS computes.*

John Searles has illustrated this difficulty (as he has for other difficulties of computationalism) by explaining that one can simulate the chemical reaction that transforms carbon dioxide into sugar, but as the information is not integrated, this simulation will not result in photosynthesis (Searle 2007). In support of this view, Prigogine insists that dissipative systems, like animal brains, survive far from thermodynamic equilibrium. As such, these systems are characterized by instability and time irreversibility in information processing. Overall, that makes organisms resistant to standard deterministic causal explanations. Instead, they can only be described statistically, in probabilistic terms, as a process whose temporal evolution is not reversible at all scales. Conversely, C.H. Bennett has shown that a Turing Machine can be made logically reversible at every step, by simply saving its intermediary results (Bennett 1973). This is generally called the irreversibility argument.[24]

Examining one aspect of this temporal irreversibility, Stephen J. Gould proposed (Gould 1989) a thought experiment that nicely illustrates the dilemma faced by those who believe that "reverse engineering" of complex biological organisms is possible through a digital deterministic platform. Gould named this the "Tape of Life Experiment" and indicated that, if a theoretical tape containing the record of all evolutionary events that led to the emergence of the human species could be rewound and then let go again, the chances that playing this tape again would generate the same sequence of events that culminated with the appearance of the human race would be equal to zero. In other words, since the tape of life would follow a path made of a huge sequence of random events, there is no hope that the precise combination that originally gave rise to mankind could be reproduced once again. Essentially, the logic behind the "Tape of

---

[24] The irreversibility argument has previously been put forward by Bringsjord and Zenzen, « Cognition is not computation, Synthese », 1997, Kluwer Academic Publishers, vol 113 issue 2 pp 285-320.

Life Experiment" strongly suggests that it is impossible to employ deterministic and reversible models to reproduce a process that emerges as a sequence of random events. Accordingly, any model running on a Turing Machine (a deterministic entity) that intends to track the evolutionary path of our species would diverge very quickly from the real process from which our species emerged. *In other words, there is no way to reverse engineer something that was never engineered in the first place.*

As a matter of fact, at the limit, the reverse engineering arguments put forward by some biologists like Craig Venter, and neurobiologists, like Henry Markram, seem to deny the quintessential role of evolution in shaping organisms in general (Venter), and the human brain (Markram) in particular. Thus, paradoxical as it may sound, the proponents of the reverse engineering view, which is considered to be at the very edge of modern biology, may not realize that by assuming this theoretical position they are frontally challenging the most enduring framework ever conceived in their field, the theory of evolution by natural selection, while favoring some sort of intelligent design scheme[25]

---

25 On further study of randomness and incomputability see Giuseppe Longo : Incomputability in Physics and Biology (Mathematical structures in computer sciences vol 12, Oct 2012, pp 880-900).

## CHAPTER 6 - The human brain as a very special physical system: The information argument

When one studies a physical system S, one's effort is usually carried out from the "outside" of S. We measure properties and the time evolution of S by exposing it to various experimental situations. But when one studies a human brain, one can obtain two different kinds of information: the one obtained from the outside, by measurements like the ones obtained from any other physical system (defined here as Type I extrinsic or Shannon-Turing information), and another one, obtained from the "inside", by questioning the owner of that particular brain (labeled here as Type II intrinsic or Gödel information). Type I information is expressed with the rigid syntax provided by numbers, bits, and bytes. Since it is generated by an integrated system (the brain), Type II information includes a rich range of semantics that amplifies the meaning and reach of the subject's language; the main way he/she communicates his/her own thoughts and feelings.

This can be illustrated by a simple experiment. When trying to find out what happens when a patient sees pictures containing upsetting images, a neuroscientist can measure some kind of electrical brain activity from the subject to obtain Shannon-Turing Type I (outside) information. On the other hand, the experimenter can simply ask the subject to express his/her feelings to gather Gödelian Type II (inside) information[26]. Once both data sets are collected, the neuroscientist then seeks to correlate the measurements with what the patient has expressed.

When we measure some pattern of electrical brain activity, it does not necessarily correlate in all circumstances to the same feeling expressed by a patient. We have already

---

[26] Gödelian information defines all information that cannot be proved to the true by a formal system used to collect information about an organism.

explained previously that this is due to various physiological properties inherent to brain, like:

(1) The initial conditions are never the same in a brain.
(2) The brain is a dynamical complex integrated system. As such, it can produce different emerging properties with immeasurably small changes in these initial conditions. Therefore, there is no hope to measure all the necessary data in real time when dealing with a living brain. Even if we were able to obtain all the required measurements, we would not necessarily know how to translate them into the subject's feelings. We would need a living brain to tell us about those.

The fact that the human brain is capable of expressing both Type I and Type II information, as well as the impossibility of finding a perfect correlation between both, poses a unique challenge for the traditional scientific approach. That is because this particular physical object we call a human brain occupies a very special position among the objects studied by natural sciences. In a brain, the outside information (digital and formal) will never be able to fully account for the whole reality depicted by the inside information (analog and integrated). It is this inside information that includes the uniqueness that emerges from the brain's amalgamation of information and matter, arguably the most powerful computational endowment bestowed to us by evolution.

Overall, the differences between Shannon-Turing and Gödelian information can be described as follows: Type I information is symbolic; that means that the recipient of a Type I message has to decode it in order to be able to extract a meaning. For this he obviously needs to know the code prior to receiving the message; if the code was included in the message it would not be accessible. Without an external code, for instance, the very lines you are reading now would have no meaning for you. Meaning is essential for your brain to be able to do something with the message. Conversely, Type II information does not need any code to be processed; its meaning is recognized instantaneously by a brain. That follows because the message's meaning is defined by the brain that generates or receives the message. That may explain, for instance, why a pair of plaques were placed on board *Pioneer 10* and *Pioneer 11* spacecraft, launched in 1972. These plates featured images, i.e. analog messages, since it was believed that these would have a better chance to be understood by some intelligent extraterrestrial beings who might come in contact with them.

As we mentioned above, in Type I information, the code cannot be embedded in the message itself. Thus, if living brains process only Type I information, they would have to have some sort of meta-code to achieve any understanding of an internal message flowing through its circuits. But to understand this meta-code, the brain would require yet another intermediary set of rules to make sense of this intermediary code. As one can easily verify, when taken to the limit such reasoning generates an infinite regress; a new code will always be needed to extract meaning from a message transmitted a level above it. That is why philosophers like Daniel Dennett describe this process as "the spectator in the Cartesian theater".

The infinite regress described above justifies our contention that brains have to be considered as integrated information systems that compute using analog components, something that is completely foreign to the way digital systems operate. This limitation also explains why computers completely fail when they are required to handle "the meaning" of things. As Marvin Minsky himself recognized[27], computers fail when faced with tasks that require common sense:

"They cannot look at a room and tell you about it".

That happens because computers can only handle Shannon-Turing information. Type II, or Gödelian information, is not their stuff. Hence, whatever immense computing power they may boast, they will fail miserably at tasks considered mundane enough for any human being.

---

[27] WIRED magazine August 2003.

Some philosophers have labeled the difficulty in correlating what we call Type I and Type II information the hard problem of neuroscience. It seems to us that this problem results, at least in part, from the fact that different human endeavors deal primarily with one of these two types of information. The method of scientific investigation naturally favors Shannon-Turing information for its reproducibility, its precision and the social consensus it can generate. Gödel's information whose communication is not precise is considered more relevant to psychology and to the arts. As such, the human brain is the only natural object we know capable of employing a "communication system" (oral and written language) to generate Gödelian Type II information to express, in an understandable way, its own inside views[28].

No Turing machine can achieve such a wondrous feat.

Brains, but not computers, can also deal with potentially conflicting or ambiguous messages carried by Type I and Type II information sampled simultaneously. To explain this point, let's return to the phantom limb phenomenon discussed in Chapter 2. It is well known that the classic Hubel and Wiesel feed-forward digital model of sensation that motivated the original proposition of the "a posteriori" binding problem" cannot account for the phenomenon of phantom limb sensation. Therefore, no digital model of the brain could yield the kind of ambiguous sensation experienced by a subject who confronts the knowledge that, even in the physical absence of his/her limb, he/she can still feel its presence. As a matter of fact, we think that the phantom limb sensation could be re-interpreted through an analogy with Gödel's first incompleteness theorem. A simple hypothetical situation can illustrate this idea. Suppose a subject who had his right arm amputated is resting in a hospital bed and cannot see his limbs because his body is fully covered with a bed sheet. Now, the physician who performed the arm amputation comes to his bedside and informs him that, regrettably, his arm had to be

---

[28] It seams to us unreasonable to consider that such an inside view can also exist in lesser complex systems, such as the ones usually studied by physics.

severed a couple of hours ago. Even though the patient now knows that this is a true fact, he experiences a profound contradiction because he still can feel the presence of his right arm, underneath the bed sheet, thanks to a clear manifestation of a phantom limb sensation. Even if the surgeon, in a tasteless act, shows the amputated arm to the patient to convince him that the surgery took place, the patient will still be unable to stop experiencing his phantom sensation. This gruesome scene illustrates that the human mind can handle cases in which provability (e.g. not having an arm any more) and feeling (e.g. not being able to deny the sensation of having an arm) diverge and coexist in the same brain. Conversely, a Turing Machine would not be able to cope with this ambiguity at all. In fact, the Turing Machine's unsuitability in dealing with such a problem would only be magnified if, as we predict, the body schema materializes as a brain-derived analogy, created by the complex interactions of NEMFs, as a result of a HDACE operation.

In the example described in the previous paragraph, intrinsic Gödelian Type II information contradicts extrinsic Shannon-Turing Type I information. As such, the existence of such a contradiction can only the experienced and reported by a subject through the expression of Gödelian, Type II information.

Obviously, all the objections raised above have been neglected or rejected by researchers who want to simulate a human brain on a digital computer. For example, Henry Markram, the main proponent of the Human Brain Project, has declared that[29]:

*"Consciousness is just a massive amount of information exchanged by trillion of brain cells ... I do not see why you should not be able to generate a conscious mind".*

In no uncertain terms this declaration completely neglects all mathematical, computational, evolutionary, and information arguments made above. According to our view, any computer

---

[29] In SEED magazine (March 2008).

simulation based only on data collected from the outside, and working within a given formal system, cannot hope to reproduce the full richness of neurological functions bred by the non-algorithmic computational machinery found in animal brains.

At this point, we should also stress that Type II information is not less "real" than Type I; it is causally effective, i.e. it induces brain changes, and it is very uniquely linked to the integrated history of each particular individual brain. Essentially, it depicts the unique point of view of each of our individual life stories and deeply influences the beliefs, thoughts, decisions and behaviors of all of us. Ultimately, it also shapes the morphology and functions of our brains. Indeed, Type II information is so real for any of us that it may explain a variety of intriguing phenomena, like the Placebo effect. Well known among health professionals, the placebo effect refers to the fact that a significant percentage of patients can perceive or even exhibit improvements in a clinical condition by taking an otherwise inert substance – like a tablet made of flour – that was labeled by a doctor as a "potential new treatment or cure" for that ailment. In other words, by simply receiving Type II information from their doctors and following the instruction to take the "miraculous flour pill", many patients may experience an improvement in their clinical state.

Since our main conjecture, as well as those of other authors, such as Penrose, Copeland, and Calude, is that biological brains cannot be reduced to a Turing Machine, is there any other view that could better describe the operation of the central nervous system of animals and explain their superior computational capacity when compared to Turing Machines?

Penrose and Hameroff's orchestrated objective reduction (Orch-OR) theory proposes that some, still unknown, quantum gravity effects should account for the way brains operate (Penrose 1994). Penrose claims that wave function collapse is a prime candidate for a non-computable process since collapse is a truly random property and, as such, it cannot be simulated. We disagree with the Orch-OR theory. Indeed, what follows is an attempt to propose an alternative theory, which does not make explicit use of quantum mechanics, but still leaves open other computational effects at different resolution levels.

As seen in Chapter 2, in total contrast to proposals that purport the brain as some sort of digital device, we introduce a new theory of brain function: the relativistic brain theory. According to this theory, internal neuronal electromagnetic fields would be responsible for creating the brain space-time continuum from where the mental space emerges. The mental space would account for all the emergent properties of the brain that define the higher order functions and behaviors produced by the central nervous system. Likewise, in the past, many theories have assigned a critical role for electromagnetic waves in the brain. A few of these theories have gone as far as to suggest that such NEMFs account for consciousness. The Conscious Electromagnetic Information (CEMI) theory, developed since 2002 by Johnjoe McFadden of the University of Surrey is clearly among the most advanced of such theories. McFadden has published a series of articles (McFadden 2002a; McFadden 2002b) detailing his theory and the findings that may support it.

The brain can certainly read Shannon-Turing information encoded in the firing rate of every single neuron and generate an output. Yet, this output alone cannot express the richness of the perceived Type II information that results from an integrated and unified entity like a human brain. Although NEMFs can be tiny when one considers a single neuron (in the range of $10^{-10}$ to $10^{-7}$ Tesla), they can become significantly stronger and more easily measurable when generated by large neuronal populations whose axons form long bundles, as part of the brain's white matter. In 2007 Thomas Radman and colleagues showed that small electric fields can have significant effects on spike timing, and hence on the neuronal coding of information (Radman, Su et al. 2007). These electrical fields also provided a precise mechanism for generating endogenous field oscillations in the gamma frequency band, which have been associated with some important brain functions, including REM sleep, a period in which memories are supposed to be consolidated at cortical level.

Radman et al. also showed that nonlinear properties of individual neurons can amplify the effects of small electrical fields by effectively resonating with them. Moreover, it has also been suggested that even very small fluctuating NEMFs could trigger or inhibit the firing of a neuron, which is close enough to its threshold, modifying how such a neuron processes a particular piece of information.

Here, we propose that NEMFs could continuously act on neural networks located far apart in the brain, building a continuous feedback loop between the two computational levels – one digital and one analog - that defines the HDACE. In our view, relying on such a hybrid digital-analog mode of operation endows a nervous system with several computational and, hence, evolutionary advantages. In computational terms, it allows the brain to constantly build an internal, continuously variable, analog computer that integrates various neural sources of information, while processing inputs from the external world much faster and with a much higher generalization capability than a digital computer. Thus, like a protein, the brain would compute by relying on the laws of physics alone, expressed in the analog domain of NEMFs, rather than on an algorithm running in a digital device.

Another advantage is that analog computing is quasi-instantaneous and does not require the type of matching precision that digital machines need to perform a pattern recognition task. Instead, analog computers build an internal analogy and compare it globally to a new incoming message, rather than decomposing the message into its building features as a digital machine does.

From an evolutionary point of view, the initial appearance of NEMFs in primitive brains, and their inherent ability to instantly process information emerging from a volume of spatially distributed neurons, may have provided an optimal way from which progressively more complex brains could emerge. Evolving larger brains, by adding more processing elements (i.e. neurons), likely required the concurrent addition of denser long-range nerve bundles, i.e. white matter, to generate more powerful NEMFs. In other words, bigger brains entail the presence of bulkier white matter because it is through these "biological coils" that strong enough NEMFs can be generated to "glue the digital component (neurons) of the brain" together. The RBT predicts that the brains of primates in general and humans in particular experienced an explosive growth of their white matter during evolution. Thus, after millions of years spent developing and refining an optimal interplay between the brain's analog and digital components, the evolutionary competitive advantage of NEMFs would have manifested itself by the progressive emergence of the kind of higher brain functions observed in higher primates, including ourselves.

In our view, time-varying NEMFs emerge from the widespread neuronal activity that replays, combines and integrates the memories stored in a distributed manner in the cortical neuronal circuits. As such, this analog component of the brain's computational engine would define an "internal analogy" that carries, embedded in it at all times, a record integrating all previous experiences of a subject from which Gödelian Type II information is generated. This "internal analogy" is manifested at each moment in time by widespread neuronal firing synchronization, across widely separated brain structures. As such, the widespread neuronal synchronization, brought about by the continuous generation of NEMFs, makes the brain operate, for all means and purposes, as a space-time continuum where widespread neuronal space is bound to time by the continuous expression of NEMFs.

The type of brain synchronization we are referring to can only be created and mediated by an analog signal (like EM fields), since a degree of imprecision is needed for the system to operate with the proper degree of adaptability. Digitally mediated synchronization, on the other hand, would not allow this process to work because of its inherent rigid precision.

According to our theory, as a "brain analogy" is built in a given moment in time, it contains, in addition to an analog computation engine, expectations manifested by a wave of anticipatory neuronal activity that creates a brain's "internal hypothesis" of what may happen in the next moment, according to the subject's previous experiences. We have named this overall anticipatory signal as the "brain's own point of view" (Nicolelis 2011).

The "brain's own point of view" refers to a kind of non-discursive, non-discrete, non-verbal, anticipatory information that is employed by the brain to build internal representations and process Gödelian information. The term relativistic brain comes from this key concept, since it portrays the notion that everything our brains do revolves around its own internal perspective and representation of material reality.

According to this relativistic view of the brain, when a new sample of the external world is captured by the multitude of peripheral body receptors (e.g. tactile, visual, auditory, etc.) and transmitted through parallel sensory pathways into the CNS, this ascending signal interferes with the "internal analogy" that defines the expectation built by the brain's analog computer engine. The resulting interferometry pattern emerging from this collision (between the peripheral signals and the expectation generated internally by the brain) gives rise to the perception of what is going around us in the surrounding world. Essentially, what we propose is that the true mechanism of reality building occurs in the analog domain, as an electromagnetic emergent property, rather than into the brain's digital component. As such, no Turing machine would be able to simulate such an internal model of reality, nor the key tasks performed by our brains that depend on this representation.

# CHAPTER 7 – Conclusions

In this monograph we raised a series of neurophysiological, mathematical, computational, evolutionary, and information arguments to support the contention that no Turing machine, no matter how sophisticated it may be, would be able to run an effective simulation of any complex animal brain, including our own. As part of our argument, we also introduced a new conjecture, named the relativistic brain theory.

Usually, the manifestations of the mental space are studied through the collection of Type II information. These manifestations include, among others, the sense of self and the "brain's own point of view", i.e. the brain's own internal account of material reality. Since in our view all behaviors generated by complex brains like ours use the "brain's own point of view" as a frame of reference, the term relativistic properly describes the central framework employed by complex central nervous systems like the human brain.

Overall, the relativistic brain theory proposes that the existence of a brain analog computational component, and its recursive interaction with digital elements, represented by the the neural networks, is what allows us to anticipate, abstract, and adapt quickly to events in the external world. In mathematical terms, the HDACE implies that complex brains generate more behaviors than those that can be computed by a Turing machine, as predicted by Gödel's first incompleteness theorem. Therefore, higher brain functions, such as creativity, intelligence, intuition, mathematical abstraction, all forms of artistic expression, empathy, altruism, fear of death, just to mention a small set of well-known human attributes, are all examples of Gödelian emergent properties of the human mind that could never be compressed by an algorithm and simulated by a digital computer. Essentially, what we intend to say is that complex and integrated brains like ours can generate innumerable emergent behaviors that far exceed the computational capability of any Turing

Machine. As such, the animal brain, in particular the human brain, could be considered as the first class of hypercomputers, devices that outperform the capabilities of the Universal Turing Machine.

Christopher Koch, chief scientific officer of the Allen Institute for Brain Science, states in his last book [30] that consciousness arises within any sufficiently complex information system (being it a brain or any other physical system). He has recently declared that:

*"Likewise, I argue that we live in a universe of space-time, mass, energy and consciousness arising out of complex systems."*

Koch refers to a function named $\phi$, imagined by Giulio Tononi (Tononi 2012), which measures in bits the amount of information generated by a complex of elements, above and beyond the information generated by its parts. Koch proposes that $\phi$ provides a measure of consciousness and that, consequently, any sufficiently complex system could be conscious, including a digital computer, leading Koch to a form of panpsychism.

We, on the other hand, believe that complexity is a necessary but not a sufficient condition to generate Gödelian information, and hence, a conscious brain. Instead, we propose that higher brain functions can only arise because of the evolutionary shaping of the physical structure of the brain that allowed information and organic matter to become intrinsically bound and intertwined. In our view, despite being an interesting concept, $\phi$ measures Shannon-Turing information and, as such, it is not sufficient to justify the emergence of higher brain functions responsible for fusing, in a single picture, a continuous stream of sensory and mnemonic information.

Instead, specific morphological and physiological brain attributes, resulting from an evolutionary struggle, are necessary for the manifestation of these higher mental functions. Therefore,

---

[30] Consciousness : Confessions of a romantic reductionist, 2012

in our opinion, it is totally nonsensical to believe that by multiplying the number of processors and increasing their memory and connectivity, higher brain functions like ours will suddenly emerge from any sophisticated version of a Turing Machine.

We should emphasize that we are not denying the fact that a digital computer, made of a large number of processors, is capable of generating emergent properties, in the shape of heat and a complex magnetic field. However, none of these properties will have anything to do with the sort of continuously changing NEMFs that a living brain is capable of generating. For once, none of the magnetic fields generated by a supercomputer are used in any meaningful computation, nor can they influence the microchips like NEMFs can influence real neurons in a recursive way.

Moreover, to simulate real NEMFs, one would have to reproduce the overall evolutionary process that, over millions of years, accounted for the shaping of all levels of organization of our present brains: from the distribution of ion channels in each of its tens of billions of neurons, to the basic macro and micro connectivity of a multitude of neuronal circuits, all the way to the maximum energy consumption limit that constrains the normal operation of the central nervous system. Indeed, researchers in the field of artificial intelligence, computer science, and computational neuroscience who claim that the fundamental operations of a complex brain can be reduced to algorithms, capable of running on a Turing machine, basically ignore all these issues. Instead, they claim that by relying on mathematical approximation they will overcome these limitations. Yet, no convincing proof of this assertion can be found in any of the recent publications from groups, like the Human Brain Project, which have based their entire research programs on the claim that digital computers will soon be able to simulate entire animal brains.

One can only suppose that the conceptual mistake of extending the Church-Turing hypothesis from the mathematical realm to physical reality has clouded the judgment of some researchers, making them ignore or neglect the essential fact that organisms are integrated entities. As such, the theoretical computational concept of an abstract mind (i.e. software), producing subtract independent information by "running" on neural circuits (i.e. hardware) does not apply to the reality of an organism's mental life. A brain does compute, but its computation is intrinsically bound to its physical structure and its does not necessarily rely on some sort of binary logic.

The HDACE model also poses a major challenge to the notion that there is a single neural code to be found in the central nervous system. Since according to this view analog neuronal fields are being generated and can influence different groups of brain cells at a given moment in time, neurons which are themselves adapting endlessly, there is no stable neural code to speak of in a relativistic brain framework. Indeed, our theory predicts that only the intrinsic dynamics of the digital-analog recursive system that defines the brain spatiotemporal continuum matters for processing information. In this context, there is no distinction between hardware and software in the HDACE model of the brain because in it a great deal of the system's hardware is computing in the analog domain, defining a fully integrated system. As such, there is no possible way to reduce the operation of the brain to an algorithm, or software, because there is none to speak of in any animal's CNS.

In such a continuous state of flux, the brain in general, and the neocortex in particular, can only be seen as a dynamic space-time continuum that processes information as a whole, and not only as a mosaic of highly specialized areas (Nicolelis 2011). That would explain why a growing number of studies has demonstrated than even primary cortical areas (e.g. somatosensory or visual cortex) are capable of representing some type of information derived from other sensory modalities. Rats that learn to perceive infrared light using their touch cortex illustrate this phenomenon well (see Chapter 1).

Furthermore, the proposition that the cortex may compute information through a space-time continuum, thanks to the "a priori binding" produced by NEMFs, indicates that classical spatial constructs, introduced in neuroscience during the past century, such as maps, cortical columns and labeled lines lose most, if not all, their meaning in terms of information processing units. In fact, considerable neurophysiological evidence already points in that direction.

We propose, instead, that widely distributed and dynamic neuronal interactions, bound in the analog domain by complex NEMFs, define a space-time neuronal continuum from which a "mental space" emerges and where all mental computations take place. In our view, such "mental space" should account for the entire realm of each subject's experiences, feelings, intellectual and cognitive faculties.

As we saw in Chapter 2, several examples illustrate well our view on how complex mental experiences can be generated, by the interaction of NEMFs that define the mental space, such as the perception of pain and the sense of self (typical Gödelian information-based phenomena). For decades neurophysiologists have attempted to locate the neuronal substrate of these brain functions by probing the functional properties of neurons located in a multitude of subcortical and cortical areas, as if these neurological phenomena were defined only by Shannon-Turing information.

Assuming that a proper mathematical language could be employed to analyze the "mental space", it is conceivable to predict that, one day in the future, such an analysis could not only be utilized to diagnose mental disorders with great precision, but also preventively detect when the "mental space" starts deviating towards a configuration that could result in neurological or psychiatric disorders much later on.

In conclusion, throughout this monograph we have argued that the "physical version" of the Church-Turing thesis and its extension to natural physical reality constitutes a serious conceptual mistake. This happens because Turing's model of computation, which includes the Turing Machine and all digital computers built upon his primordial model, does not necessarily provide a comprehensive picture of the computational power of natural physical objects, particularly integrated systems known as animal brains, the most powerful class of computational devices produced by evolution as far as one can tell. In this context, the relativistic brain theory supports the possibility that effective hypercomputers can exist in the physical world. They are called brains. And, according to the vast list of arguments listed above, their computational capabilities far exceed that of any Turing machine and its derivatives.

As we saw above, to some degree, brain-machine interfaces (BMIs), which connect the brains of animals to artificial actuators via a computational interface, represent another type of artificial or hybrid hypercomputer. Recent advances in the BMI field, such as brain-to-brain interfaces (Pais-Vieira, Lebedev et al. 2013), a paradigm in which multiple animal brains can exchange information seamlessly in order to interact directly to perform a particular task, like control of an artificial, real or virtual actuator (Ramakrishnan, Ifft et al. 2015), show great potential in expanding research on designing and testing a biologically-inspired hypercomputer.

It is important to emphasize that the relativistic brain theory also addresses a series of classical philosophical questions regarding the hypothesis that a complete account of all brain functions can be given in strictly measurable neurophysiological terms. This view is labeled the monistic materialistic paradigm. It proposes that there is nothing more to know about the whole brain and its functions than what can be inferred by measuring material behaviors and thus denying the difference between Type I and Type II information streams. This paradigm opposes the Cartesian dualistic approach where brain and mind are considered as separated entities. Dualism is considered as non-scientific, since it involves an abstract non-material entity (the mind) that is able to act on matter (the brain).

From our viewpoint, we cannot see how an abstract non-material entity could exchange energy in order to act on a material object. Therefore, the relativistic brain theory and its HDACE remains a monist materialistic view, although it introduces and justifies the concept of a "mental space" that takes the shape of emergent NEMFs. These NEMFs are able to compute and to modify the very material substrate, i.e. the neural networks, from which it emerged, reestablishing a peculiar sort of dualism in which information measured from outside and information felt from inside are not isomorphic and, thus, justifying why all brain functions are indeed not directly measurable. The relativistic brain approach, although it remains physicalist and monistic, considers that all information necessary to account for a living brain cannot only be collected as Shannon-Turing information from the outside.

Many authors, including Turing, Gödel, Chaitin, Copeland and Roger Penrose are convinced, as we are, that the human brain far exceeds the capabilities of any possible Turing machine. Some of these authors, however, have been looking for some source of a "non-determinist" neural substrate that could account for the emergence of attributes of the human nature, things such as creativity and free will, and be used to distinguish us from deterministic automatons. The relativistic brain model suggests such a procedure without, as Penrose does, necessarily having to consider the explicit harnessing of quantum phenomena by the CNS[31]. Instead, we argue that the sensitivity to initial conditions of a relativistic brain and the dynamical complexity and non-computability of the HDACE provide the neuronal substrate for probabilistic higher order brain functions, including creativity and free will, to emerge, although they are obviously neither measurable from outside, nor reproducible by a deterministic formal system. Accordingly, we propose that human life experiences, created and recorded by our brains, are truly unique natural phenomena, which will forever remain far beyond the reach of any Turing machine.

Finally, we would like to state that, by no means, are we suggesting that higher brain functions will never be reproduced by artificial means. Essentially, we simply put forward a series of arguments that demonstrate that if and when that happens, it will

---

[31] The relativistic brain theory does not necessarily exclude the existence of quantum phenomena in the brain.

not be via a Turing machine, no matter how powerful and sophisticated it is.

By the same token, we have no intention to diminish the significant services that Turing Machines and artificial intelligence may provide to mankind and the progress of society. However, we feel compelled to make the point that, in a moment in which the scientific community and governments try to concentrate their collective efforts to reach a deeper understanding of the human brain, it would be a tragedy to waste effort, careers, and limited resources by investing in theoretically flawed research programs that, at their very core covertly deny the role of evolution in shaping organisms, while bluntly sustaining that brains are nothing but the product of sophisticated electronic versions of steam machines.

# Bibliography

## 1. LISTENING TO NEURONS AND BUILDING BMIS

Carmena, J. M., M. A. Lebedev, et al. (2003). "Learning to control a brain-machine interface for reaching and grasping by primates." PLoS Biol 1(2): E42.

Chapin, J. K., K. A. Moxon, et al. (1999). "Real-time control of a robot arm using simultaneously recorded neurons in the motor cortex." Nat Neurosci 2(7): 664-670.

Hebb, D. O. (1949). The organization of behavior; a neuropsychological theory. New York,, Wiley.

Nicolelis, M. (2011). Beyond boundaries : the new neuroscience of connecting brains with machines--and how it will change our lives. New York, Times Books/Henry Holt and Co.

Nicolelis, M. A. (2001). "Actions from thoughts." Nature 409(6818): 403-407.

Nicolelis, M. A. (2003). "Brain-machine interfaces to restore motor function and probe neural circuits." Nat Rev Neurosci 4(5): 417-422.

Nicolelis, M. A. and J. K. Chapin (2002). "Controlling robots with the mind." Sci Am 287(4): 46-53.

Nicolelis, M. A. and M. A. Lebedev (2009). "Principles of neural ensemble physiology underlying the operation of brain-machine interfaces." Nat Rev Neurosci 10(7): 530-540.

Nicolelis, M. A. L. (2008). Methods for neural ensemble recordings. Boca Raton, CRC Press.

Patil, P. G., J. M. Carmena, et al. (2004). "Ensemble recordings of human subcortical neurons as a source of motor control signals for a brain-machine interface." Neurosurgery 55(1): 27-35; discussion 35-28.

Schwarz, D. A., M. A. Lebedev, et al. (2014). "Chronic, wireless recordings of large-scale brain activity in freely moving rhesus monkeys." Nat Methods 11(6): 670-676.

Thomson, E. E., R. Carra, et al. (2013). "Perceiving invisible light through a somatosensory cortical prosthesis." Nat Commun 4: 1482.

Wessberg, J., C. R. Stambaugh, et al. (2000). "Real-time prediction of hand trajectory by ensembles of cortical neurons in primates." Nature 408(6810): 361-365.

## 2. THE RELATIVISTIC BRAIN THEORY PRINCIPLES

Anastassiou, C. A., S. M. Montgomery, et al. (2010). "The effect of spatially inhomogeneous extracellular electric fields on neurons." J Neurosci 30(5): 1925-1936.

Arvanitaki, A. (1942). "Effects evoked in an axon by the activity of a contiguous one." J. Neurophysiol. 5: 89-108.

Bakhtiari, R., N. R. Zurcher, et al. (2012). "Differences in white matter reflect atypical developmental trajectory in autism: A Tract-based Spatial Statistics study." Neuroimage Clin 1(1): 48-56.

Berger, H. (1929). "Electroencephalogram in humans." Archiv Fur Psychiatrie Und Nervenkrankheiten 87: 527-570.

Botvinick, M. and J. Cohen (1998). "Rubber hands 'feel' touch that eyes see." Nature 391(6669): 756.

Cohen, L. G., P. Celnik, et al. (1997). "Functional relevance of cross-modal plasticity in blind humans." Nature 389(6647): 180-183.

Copeland, B. J. (1998). "Turing's O-machines, Searle, Penrose and the brain (Human mentality and computation)." Analysis 58(2): 128-138.

Debener, S., M. Ullsperger, et al. (2005). "Trial-by-trial coupling of concurrent electroencephalogram and functional magnetic resonance imaging identifies the dynamics of performance monitoring." J Neurosci 25(50): 11730-11737.

Engel, A. K., P. Fries, et al. (2001). "Dynamic predictions: oscillations and synchrony in top-down processing." Nat Rev Neurosci 2(10): 704-716.

Englander, Z. A., C. E. Pizoli, et al. (2013). "Diffuse reduction of white matter connectivity in cerebral palsy with specific vulnerability of long range fiber tracts." Neuroimage Clin 2: 440-447.

Fingelkurts, A. A. (2006). "Timing in cognition and EEG brain dynamics: discreteness versus continuity." Cogn Process 7(3): 135-162.

Fuentes, R., P. Petersson, et al. (2009). "Spinal cord stimulation restores locomotion in animal models of Parkinson's disease." Science 323(5921): 1578-1582.

Ghazanfar, A. A. and C. E. Schroeder (2006). "Is neocortex essentially multisensory?" Trends Cogn Sci 10(6): 278-285.

Gray, J. (2004). Consciousness: creeping up on the hard problem. USA, Oxford University Press

Hubel, D. H. (1995). Eye, brain, and vision. New York, Scientific American Library : Distributed by W.H. Freeman Co.

Jefferys, J. G. (1995). "Nonsynaptic modulation of neuronal activity in the brain: electric currents and extracellular ions." Physiol Rev 75(4): 689-723.

Jibu, M. and K. Yasue (1995). Quantum brain dynamics and consciousness : an introduction. Amsterdam ; Philadelphia, J. Benjamins Pub. Co.

John, E. R. (2001). "A field theory of consciousness." Conscious Cogn 10(2): 184-213.

Kreiter, A. K. and W. Singer (1996). "Stimulus-dependent synchronization of neuronal responses in the visual cortex of the awake macaque monkey." J Neurosci 16(7): 2381-2396.

Kupers, R., M. Pappens, et al. (2007). "rTMS of the occipital cortex abolishes Braille reading and repetition priming in blind subjects." Neurology 68(9): 691-693.

McFadden, J. (2002a). "The conscious electromagnetic information (Cemi) field theory - The hard problem made easy?" Journal of Consciousness Studies 9(8): 45-60.

McFadden, J. (2002b). "Synchronous firing and its influence on the brain's electromagnetic field - Evidence for an electromagnetic field theory of consciousness." Journal of Consciousness Studies 9(4): 23-50.

Melzack, R. (1973). The puzzle of pain. New York,, Basic Books.

Melzack, R. (1999). "From the gate to the neuromatrix." Pain Suppl 6: S121-126.

Nicolelis, M. (2011). Beyond boundaries: the new neuroscience of connecting brains with machines--and how it will change our lives. New York, Times Books/Henry Holt and Co.

Nicolelis, M. A., L. A. Baccala, et al. (1995). "Sensorimotor encoding by synchronous neural ensemble activity at multiple levels of the somatosensory system." Science 268(5215): 1353-1358.

O'Doherty, J. E., M. A. Lebedev, et al. (2011). "Active tactile exploration using a brain-machine-brain interface." Nature 479(7372): 228-231.

Pais-Vieira, M., M. A. Lebedev, et al. (2013). "Simultaneous top-down modulation of the primary somatosensory cortex and thalamic nuclei during active tactile discrimination." J Neurosci 33(9): 4076-4093.

Papanicolaou, A. C. (2009). Clinical magnetoencephalography and magnetic source imaging. Cambridge, UK ; New York, Cambridge University Press.

Papoušek , H. and M. Papoušek (1974). "Mirror image and self-recognition in young human infants: I. A new method of experimental analysis." Developmental Psychobiology 7(2): 149-157.

Pockett, S. (2000). The Nature of Conciousness: A Hypothesis. Lincoln, NE, iUniverse.

Ribeiro, S., D. Gervasoni, et al. (2004). "Long-lasting novelty-induced neuronal reverberation during slow-wave sleep in multiple forebrain areas." PLoS Biol 2(1): E24.

Sadato, N., A. Pascual-Leone, et al. (1996). "Activation of the primary visual cortex by Braille reading in blind subjects." Nature 380(6574): 526-528.

Santana, M. B., P. Halje, et al. (2014). "Spinal cord stimulation alleviates motor deficits in a primate model of Parkinson disease." Neuron 84(4): 716-722.

Shokur, S., J. E. O'Doherty, et al. (2013). "Expanding the primate body schema in sensorimotor cortex by virtual touches of an avatar." Proc Natl Acad Sci U S A 110(37): 15121-15126.

Tsakiris, M., M. Costantini, et al. (2008). "The role of the right temporo-parietal junction in maintaining a coherent sense of one's body." Neuropsychologia 46(12): 3014-3018.

Uttal, W. R. (2005). Neural theories of mind: why the mind-brain problem may never be solved. Mahwah, N.J., Lawrence Erlbaum Associates.

von der Malsburg, C. (1995). "Binding in models of perception and brain function." Curr Opin Neurobiol 5(4): 520-526.

Yadav, A. P., R. Fuentes, et al. (2014). "Chronic spinal cord electrical stimulation protects against 6-hydroxydopamine lesions." Sci Rep 4: 3839.

## 3. THE MISMATCH BETWEEN BRAINS AND TURING MACHINES

Copeland, B. J. (2002). "Hypercomputation." Minds and Machines 12(4): 461-502.

Copeland, B. J., C. J. Posy, et al., Eds. (2013). Computability: Turing, Gödel, and beyond. Cambridge (MA), MIT Press.

Fodor, J. (1975). The language of thought. Cambridge (MA), MIT Press.

Kurzweil, R. (2005). The singularity is near : when humans transcend biology. New York, Viking.

Mitchell, M. (2009). Complexity: A guided tour. Oxford, Oxford University Press.

Poincaré, H. (1902). La science e l'hypothèse. Paris, Flamarion.

Putnam, H. (1979). Mathematics, matter, and method. Cambridge, Cambridge University Press.

Turing, A. M. (1936). "On Computable Numbers, with an Application to the Entscheidungsproblem." Proc. London Math. Soc. s2 - 42 (1): 230-265.

## 4. THE MATHEMATICAL AND COMPUTATIONAL ARGUMENTS

Anfinsen, C. B. (1973). "Principles that govern the folding of protein chains." Science 181(4096): 223-230.

Bailly, F. and G. Longo (2011). Mathematics and the natural sciences. The physical singularity of life. London, Imperial College Press.

Bentley, P. J. (2009). "Methods for improving simulations of biological systems: systemic computation and fractal proteins." J R Soc Interface 6 Suppl 4: S451-466.

Chaitin, G., N. da Costa, et al. (2011). Goedel's Way: Exploits into an undecided world, CRC Press.

Deutsch, D. (1997). The fabric of reality. Harmondsworth, Allen Lane, The Penguin Press.

Poincaré, H. (1905). Leçons de mécanique celeste. Paris, Gauthier-Villars.

Pour-El, M. B. and J. I. Richards (1989). Computability in analysis and physics. Berlin, Springer-Verlag.

Prigogine, I. (1996). The end of certainty. New York, The Free Press.

Reimann, M. W., C. A. Anastassiou, et al. (2013). "A biophysically detailed model of neocortical local field potentials predicts the critical role of active membrane currents." Neuron 79(2): 375-390.

Turing, A. M. (1936). "On Computable Numbers, with an Application to the Entscheidungsproblem." Proc. London Math. Soc. s2 - 42 (1): 230-265.

Turing, A. M. (1939). Systems of logic based on ordinals Ph. D., Princeton university.

Turing, A. M. (1946). ACE machine project. Report to the National Physical Laboratory Executive Committee.

Turing, A. M. (1950). "Computing machinery and intelligence." Mind: 433-460.

5. THE EVOLUTIONARY ARGUMENT

Bennett, C. H. (1973). "Logical reversibility of computation." IBM Journal of Research and Development 17(6): 525-532.
Bringsjord, S. and K. Arkoudas (2004). "The modal argument for hypercomputing minds." Theoretical Computer Science 317(1-3): 167-190.
Copeland, B. J., C. J. Posy, et al., Eds. (2013). Computability: Turing, Gödel, and beyond. Cambridge (MA), MIT Press.
Gödel, K. (1931). "Über formal unentscheidbare Sätze der Principia Mathematica und verwandter Systeme 1." Monatshefte für Mathematik und Physik 38: 173-198.
Gould, S. J. (1989). Wonderful life : the Burgess Shale and the nature of history. New York, W.W. Norton.
Lucas, J. R. (1961). "Minds, Machines and Gödel." Philosophy 36(112-127): 43-59.
Penrose, R. (1991). The emperor's new mind : concerning computers, minds, and the laws of physics. New York, N.Y., U.S.A., Penguin Books.
Prigogine, I. (1996). The end of certainty. New York, The Free Press.
Schrödinger, E. (1944). What Is Life? The Physical Aspect of the Living Cell Cambridge University Press
Searle, J. R. (2007). Freedom and neurobiology. New York, Columbia University Press.

6. THE HUMAN BRAIN AS A VERY SPECIAL PHYSICAL SYSTEM

McFadden, J. (2002a). "The conscious electromagnetic information (Cemi) field theory - The hard problem made easy?" Journal of Consciousness Studies 9(8): 45-60.
McFadden, J. (2002b). "Synchronous firing and its influence on the brain's electromagnetic field - Evidence for an electromagnetic field theory of consciousness." Journal of Consciousness Studies 9(4): 23-50.

Nicolelis, M. (2011). Beyond boundaries : the new neuroscience of connecting brains with machines--and how it will change our lives. New York, Times Books/Henry Holt and Co.

Penrose, R. (1994). Shadows of the mind : a search for the missing science of consciousness. Oxford ; New York, Oxford University Press.

Radman, T., Y. Su, et al. (2007). "Spike timing amplifies the effect of electric fields on neurons: implications for endogenous field effects." J Neurosci 27(11): 3030-3036.

## 7. CONCLUSIONS

Nicolelis, M. (2011). Beyond boundaries : the new neuroscience of connecting brains with machines--and how it will change our lives. New York, Times Books/Henry Holt and Co.

Pais-Vieira, M., M. Lebedev, et al. (2013). "A brain-to-brain interface for real-time sharing of sensorimotor information." Sci Rep 3: 1319.

Ramakrishnan, A., P. J. Ifft, et al. (2015). "Computing Arm Movements with a Monkey Brainet." Sci Rep, In press.

Tononi, G. (2012). Phi: A voyage from the brain to the soul. Singapore, Pantheon Books.

## APPENDIX I

### Predictions of the Relativistic Brain Theory

1) No digital simulation in a Turing machine, no matter how sophisticated it is, can simulate the complexity of the mammalian central nervous system.

2) Shannon information is not sufficient to account for or quantify all information and knowledge generated by the human brain. That is because Shannon information is purely syntactic and cannot convey the richness of semantics and ambiguity that characterizes human behavior. That can only be expressed through Type II or Gödelian information.

3) Since higher order animal brains cannot be replicated by a Turing machine, evolution by natural selection gives rise to hypercomputers, i.e. complex systems capable of outperforming a universal Turing machine. That suggests that hypercomputers can only emerge through evolution and not by mechanical construction.

4) Widely distributed neuronal ensembles define the true functional unit of the mammalian central nervous system. No single neuron is capable of sustaining a complex behavior.

5) There should be measureable, albeit small, electromagnetic fields originating from the main bundles of the diencephalic white matter that are normally not captured by traditional magnetoencephalography. Bundles of white matter would work as true "biological coils."

6) Cortical and subcortical NEMFs should be capable of inducing widespread synchronous neuronal firing across most of the brain.

7) Despite being rather small, cortical and subcortical NEMFs should be strong enough to "bind" neuronal groups located far apart in the cortex prior to the discrimination of a sensory stimulus. That should be reflected by the observation of widespread synchronous anticipatory modulation in neuronal firing rate across most of the neocortex.

8) Disruption of these NEMFs should lead to disruption in the subject's perceptual, motor, or cognitive functions.

9) At different trials of the same behavioral tasks, different NEMFs will precede the animal behavioral response. In other words, different NEMFs can generate the same behavioral outcome.

10) Different groups of neurons can cooperate to generate a similar NEMF. That means that the precise neuronal origin of a NEMF can never be anticipated.

11) Since the RBT assumes that the cortex works as a space-time continuum, multimodal sensory evoked responses could be observed across all primary sensory cortices.

12) According to the RBT, the brain continuously checks the validity of its internal model of reality. As such, sensory cortical evoked responses emerge from the interference of an ascending sensory signal and the internal dynamic state of the brain. The same physical stimulus can generate very different cortical evoked sensory responses according to the animal behavioral state. For instance, the same tactile stimulus should produce completely different cortical evoked responses in anesthetized versus fully awake and mobile animals.

13) Higher order phenomena like pain perception, phantom limb sensation and other tactile, auditory and visual illusions, visual "filling in", dreaming, and the perception of one's

sense of self are all manifestations of analog brain processing. Therefore, manipulations of NEMF could induce or interfere with such phenomena.

14) Dynamics and plasticity are the key elements governing brain operation. Therefore, the RBT predicts that no fixed neuronal code can be identified.

15) Strict feed-forward processing, topographic maps, neuronal modules, cortical columns and all other purely spatial constructs only account for neurophysiological properties observed in low dimension brain states (e.g. deep anesthesia). As such, they provide a very misleading description of brain function. Therefore, in freely behaving animals, such constructs cannot account for the large richness of neurophysiological dynamics. The RBT predicts that during anesthesia, coma or deep sleep, the white matter "biological coils" are not engaged. As a consequence, full wakefulness and awareness requires the full recruiting of the white matter core.

16) The RBT predicts that as evolution progressed and the complexity of animal behaviors increased, the contribution of white matter to the total brain mass had to increase significantly. Such an increase in white matter mass should have accelerated even more with the emergence of primates and humans.

17) The geometry and topology of the "mental space," which is defined by a combination of NEMFs that underlies a continuum multi-dimensional manifold, could be, in theory, investigated formally by non-Euclidean geometry.

18) According to the RBT, neurological and psychiatric disorders are manifestations of particular types of folding of the neuronal space-time continuum. As such, they could be

treated by therapies that manipulate the NEMFs in order to correct the erroneous folding of the space-time continuum.

19) Some forms of autism could result from the lack of proper development of white matter "biological coils." As such, these patients should exhibit abnormal folding of the cortical space-time continuum, reflected by peculiar NEMFs. Non-invasive transcranial electromagnetotherapy may serve as a treatment for such a condition.

20) Artificial tools, such as prosthetic arms and legs or even virtual bodies, could be assimilated as an extension of the subject's body schema into the brain. Neuronal space would be dedicated to represent these new artificial "body parts."

21) According to the RBT, subjects could acquire new sensory modalities that endow them with the ability to perceive new physical stimuli. That would happen by mixing the representation of the new physical dimension with an existing one, i.e. a representation of infrared light on the primary somatosensory cortex.

22) At the limit, mechanization of processes reduces the diversity and expression of natural human behaviors. Reduction of human behavior diversity diminishes human mental capacity.

23) The Singularity hypothesis, as proposed by Ray Kurzweil, is a mathematical impossibility. Intelligent machines will not be able to mimic human intelligence nor will they overcome human nature.

Made in the USA
Middletown, DE
27 May 2015